Managing Agricultural Landscapes for Environmental Quality

Strengthening the Science Base

Max Schnepf and Craig Cox, Editors

Managing Agricultural Landscapes for Environmental Quality

Strengthening the Science Base

Max Schnepf and Craig Cox, Editors

Soil and Water Conservation Society
Ankeny, Iowa

Soil and Water Conservation Society
945 SW Ankeny Road, Ankeny, IA 50023
www.swcs.org

5 4 3 2 1
ISBN 978-0-9769432-4-2

Library of Congress Cataloging-in-Publication Data

Managing agricultural landscapes for environmental quality :
strengthening the science base / Max Schnepf and Craig Cox, editors.
    p. cm.
Includes index.

ISBN 978-0-9769432-4-2

1. Agriculture--Environmental aspects.   2. Agricultural landscape management.   3. Environmental impact
analysis.   I. Schnepf, Max, 1941-   II. Cox, Craig A. (Craig Alan)

S589.75.M358 2007
333.76′16—dc22

2007012192

The Soil and Water Conservation Society (SWCS) is a nonprofit scientific and educational organization that
serves as an advocate for natural resource professionals and for science-based conservation policy. SWCS
fosters the science and art of soil, water, and environmental management on working lands to achieve
sustainability. SWCS members promote and practice an ethic that recognizes the interdependence of people
and their environment.

Contents

Foreword

Craig Cox

The "Managing Agricultural Landscapes for Environmental Quality: Strengthening the Science Base" workshop was one of a series of activities the Soil and Water Conservation Society (SWCS) has undertaken in support of the U.S. Department of Agriculture's Conservation Effects Assessment Project (CEAP). The purpose of CEAP can be narrowly defined as an effort to improve our ability to quantify the environmental effects of conservation practices applied to agricultural land. That narrow definition, in my opinion, misses the most important and lasting contribution of CEAP—building the science base for effective, sustained, and confident management of agricultural landscapes to improve soil, water, air, and fish and wildlife habitat. The reason we need to better quantify the environmental effects of conservation practices, in other words, is to better focus our conservation programs and activities where they will do the most good.

The workshop that is the subject of this report grew out of an ambitious effort, led by Max Schnepf, to document the scientific knowledge of the effects of conservation practices on the environment. The book *Environmental Benefits of Conservation on Cropland: The Status of Our Knowledge* recently published by SWCS is the product of that effort. As Max notes in his preface to that book, the literature review stopped at the edge of the

crop field—the stopping point of most research on the effects of conservation practices in recent decades. The limitations of this edge-of-field focus were evident early on to everyone involved in the "Environmental Benefits of Conservation on Cropland" project. Indeed, participants at a workshop organized early in the project strongly recommended that SWCS take on the task of pulling together the state of our knowledge of the effects of conservation practices at watershed and landscape scales. The "Managing Agricultural Landscapes for Environmental Quality" workshop is our first attempt to take on that task.

I have been privileged to participate in, and in some cases lead, the activities SWCS has undertaken in support of CEAP. I have learned a great deal in the process. The most important lesson, I think, is the essential difference between an environmental effect and an environmental benefit. We can document many environmental effects of conservation practices at the field or farm level, but those effects don't produce meaningful benefits until they are expressed at the watershed or landscape scale. Environmental quality is an aggregate phenomenon. It is the result of multiple activities—often highly disproportionate in their individual contribution—that add up to a perceptible improvement in a component of the environment important to local, national, or global communities. The effect of conservation tillage on reduced sediment loads doesn't become a benefit, in other words, until the aggregate effect is enough to reduce the number of beach closings or to increase the viability of natural reproduction of trout.

The most important contribution of science and professional judgment in conservation is, I think, to connect the dots. Conservationists must understand how individual effort at the farm and ranch level adds up to real and meaningful results at the watershed or landscape scale. Conservationists must effectively direct conservation effort based on that understanding. Unless we learn how to connect the dots at the appropriate watershed or landscape scale, our efforts at the farm or ranch scale will come to naught.

We undertook this project in a spirit of exploration and experiment. We knew that many people have struggled and continue to struggle with the challenge of understanding the benefit of conservation at the watershed and landscape scale. Our plan was to contribute to this effort by bringing together a multidisciplinary group of scientists and practitioners to share their understanding of how to connect the dots. The result far exceeded my expectations. I, for one, left the workshop with more confidence that we can connect the dots. I also left the workshop with greater conviction that connecting the dots will dramatically improve the effectiveness of our conservation efforts.

Confidence and conviction alone will not build the science base we need or translate that science base into practice. We have a great deal of work to do. I hope our workshop and this book contribute to that important work.

Preface

Max Schnepf

In January 2007, the Soil and Water Conservation Society (SWCS) completed work on a book summarizing what the scientific literature tells us about the environmental effects of applying individual conservation practices on cropland. *Environmental Benefits of Conservation on Cropland: The Status of Our Knowledge* was assembled at the request of U.S. Department of Agriculture (USDA) officials involved in the Conservation Effects Assessment Project (CEAP) initiative. Work on that book in late 2005 and early 2006 prompted discussion about what could be done by conservation researchers to extend the effort to quantify the environmental benefits of applying conservation practices on agricultural land beyond the field and farm scales—the focal point of most research in recent decades.

Subsequent discussion with USDA and other scientists led to the suggestion that SWCS plan a workshop on how to strengthen the science and account for the environmental benefits of conservation efforts on agricultural land at landscape and watershed scales. During the course of those discussions, which took place over several months' time, four recurring themes emerged. Those four themes ultimately provided the framework for an international workshop— "Managing Agricultural Landscapes for Environmental Quality, Strengthening the Science Base"—

that was held October 11–13, 2006, in Kansas City, Missouri. The themes were as follows:

1. What should we measure, and how, to account for environmental effects at landscape and watershed scales?
2. Methods for environmental management research at landscape and watershed scales.
3. The science of targeting within landscapes and watersheds to improve conservation effectiveness.
4. Realistic expectations about the timing between conservation implementation and environmental effects—lessons learned from long-term research (or what scale of change is possible over what period of time).

From the outset, the emphasis was on planning a multidisciplinary workshop in an attempt to assess what science has to tell us about applying conservation more effectively and efficiently on a landscape or watershed scale—beyond the usual research plot, field, or farm scale—and what the implications of that might be for strengthening the science through research and extension in the future. A secondary objective of the workshop was to expose the scientific community to the needs and expectations of conservation policymakers and practitioners struggling to account for the environmental benefits of projects intended to enhance soil, water, air, and habitat at broader geographic scales.

To set the stage for discussion at the workshop, SWCS commissioned the preparation of a paper for presentation at the workshop on each of the four aforementioned themes. A lead author was invited to write each paper in collaboration with a multidisciplinary team of physical and social scientists. Two additional individuals were then invited to present "value-added" perspectives on each of the four themes at the workshop. Drafts of the four commissioned papers were shared with these eight perspective presenters in advance of the workshop.

To extend the discussion about strengthening the science of conservation at landscape and watershed scales during the workshop, SWCS also issued a call for oral papers and poster papers. The emphasis in this call for papers was on research and experience at landscape and watershed scales or work that had application at these scales. More than 250 proposals were ultimately submitted. About 110 of these were selected for oral presentation; another 80 were chosen as posters.

This book contains the four commissioned papers, along with the two perspective presentations that accompanied each (save for one paper in the research methods segment that was not available for publication). A fifth part of the book features comments offered by conservation policymakers and practitioners—users of research—on what they would like in the way of new information from the research community.

A short "Roundtable" section follows each of the four commissioned papers in this book. These sections summarize four roundtable discussions held one evening during the workshop. The roundtables involved the speakers from each of the four themed sessions and workshop participants, who attended the roundtable of their choice. In addition to giving speakers and workshop participants an opportunity to elaborate on what was said during the formal sessions at the workshop, roundtable participants were asked to identify what they considered to be the five most important "next steps" that might be taken to strengthen the science surrounding each particular theme.

Prior to the workshop, a decision was made not to publish the oral and poster papers from the workshop's technical sessions in this book, but rather post those made available by the presenters on the SWCS website after the fact. Most of those presentations are available at www.swcs.org. The workshop program also is available at this site.

Any number of individuals and institutions deserve thanks for their contributions to and support of this event. Nine governmental agencies supported the workshop and publication of this book financially: Agricultural Research Service; Agriculture and Agri-Food Canada; Cooperative State Research, Education and Extension Service; Economic Research Service; Farm Service Agency; Natural Resources Conservation Service; U.S. Fish and Wildlife Service; U.S. Forest Service; and U.S. Geological Survey. The International Cooperation Office of the Mexico National Institute for Forestry, Agriculture and Animal Husbandry Research lent its name to the list of workshop sponsors. Dozens of individuals served on the technical committee that laid the groundwork for the workshop, on the program committee that helped establish the workshop format, and on the abstract review committee. Many of these same individuals and others served as session moderators during the workshop. Special thanks are due Mike Burkart and Cathy Kling, Iowa State University, who helped SWCS staff members formulate an initial plan for the workshop and Peter Groffman, Institute of Ecosystem Studies, who served as chair of the program committee.

Many farmers and ranchers work hard to conserve soil, water, air, and fish and wildlife resources within the confines of their individual agricultural enterprises. But many of society's environmental management goals require the collective action of producers across multiple farm or ranch units to achieve the critical mass of conservation action needed to make a difference in soil, water, and air quality; water conservation; and improved habitat for both fish and wildlife. This fact presents a significant challenge to researchers who must think about what research they conduct, and how, to help conservation policymakers and practitioners better serve their constituents and the environment. Likewise, this fact challenges conservation policymakers and practitioners to demand more of the research community in the ongoing effort to improve the efficacy and efficiency of conservation policy and programs. Hopefully, the Kansas City workshop moved thinking in each of these communities a step beyond where it has been.

Part **1**

What should
we measure,
and how,
to account
for environmental
effects?

Ecosystem services in agricultural landscapes

Peter Groffman
Paul Capel
Kurt Riitters
Wanhong Yang

Concepts from ecosystem services analysis, landscape ecology, and adaptive management enable the development of ideas about how to answer the question of "what to measure, and how, to account for environmental effects" in agricultural landscapes. The ecosystem services approach provides a basis for developing clear statements about the objectives to be achieved in balancing agricultural production and environmental quality. Concepts from landscape ecology offer guidance for development of monitoring and modeling to account for spatial scale and interactions between and among production and nonproduction activities. Adaptive management helps to develop an iterative process in which monitoring results are used to improve models and management over time.

Three specific questions must be addressed to determine what to measure, and how, to account for environmental effects in agricultural landscapes:

1. What indicators should be measured that are tied explicitly to the environmental goods and services supplied by agricultural landscapes—the goods and services that decision-makers and the public value and identify with?

2. What is the appropriate scale at which to measure those indicators? The relevant landscape depends upon the service (and, therefore, the measure) selected. For example, if we care about drinking water, then the scale of measurement may be the point where water is withdrawn. If we care about endangered species, then the scale may be a multistate ecological region.

3. What is the appropriate role for simulating environmental effects using models instead of, or in addition to, monitoring to quantify these effects?

These are critically important questions. There is great interest in improving the environmental performance of agricultural systems. Many resources have been allocated to this end. But results have been mixed, with continued degradation of soil, air, water, and ecological resources across the country.

The lack of tools to assess just which measures work, and why, is a critical constraint to developing and implementing improved agricultural practices. Concepts from ecosystem services and landscape ecology can help develop those tools and, more fundamentally, implement a robust, iterative adaptive management process where goals are set, actions are taken, and monitoring and modeling are used to assess outcomes. Then goals are reevaluated, practices are changed, and monitoring and modeling go on to assess outcomes. Such a process can catalyze an ongoing, evolutionary program of basic and applied research and assessment that should lead to real improvements in environmental quality on agricultural landscapes.

Ecosystem services

Ecosystem services are defined as "flows of materials, energy, and information from natural capital stocks that are combined with manufactured and human capital services to produce human welfare" (Daily et al., 1997). The concept was developed to put economic values on the nonmarket goods and services provided by ecosystems that are often damaged by agricultural and industrial activities (Carpenter and Turner, 2000). Costanza et al. (1997) estimated that 17 ecosystem services for 16 biomes provide an average $33 trillion per year of value. In comparison, the U.S. Bureau of Economic Analysis estimate of the gross domestic product (a measure of the market value of goods, services, and structures produced in the economy) was approximately $12.5 trillion in 2005 dollars (source: Table 9 in https://bea.gov/bea/newsrel/gdpnewsrelease.htm). Ecosystem services may rival nonecosystem services, but quantification of ecosystem services may require more experience before such comparisons can be drawn.

In addition to providing a means for comprehensive quantification of environmental damage, the ecosystem services concept is useful for evaluating tradeoffs and conflicts between different goals and values (Rodriquez et al., 2006). An ecosystem services-based approach allows for consideration of a wide range of variables, from production to water quality, to aesthetic considerations, to the presence of rare species (Groffman et al., 2004). Tradeoffs and conflicts are a hallmark of efforts to manage agricultural landscapes to balance production of food and fiber with environmental quality (Table 1).

Tradeoffs between and among ecosystem services are made in both the spatial and temporal domains. Use of fertilizer in one area that results in a decline in water quality in another is an example of a spatial tradeoff, while agricultural production that results in salinization of irrigated soils is an example of a temporal tradeoff (Rodriquez et al., 2006). Reconciling these tradeoffs requires classifying and placing value on the different services, for example, "provisioning" services, such as agricultural production; "cultural" services, such as the aesthetics of landscape views; and "supporting" services, such as water purification (DeFries et al., 2004).

Many aspects of an ecosystem services approach have already been applied in agricultural production systems. The economic production from agricultural landscapes has always been analyzed at scales ranging from individual fields to the nation as a whole, and individuals have always been concerned about the health of particular land parcels insofar as it affects commodity production. More recently, research to identify indicators for monitoring and assessing ecological aspects of agricultural landscapes has identified many conceptual and operational approaches.

For example, Freyenberger et al. (1997) and van der Werf and Petit (2002) considered whole-farm approaches, while others focused on specific endpoints, such as biodiversity (e.g., Clergue et al., 2005), water quality (e.g., Schröder et al., 2004), and soil quality (e.g., Bouma, 2002) at local to regional scales. Implementation of agricultural indicators in broad-scale environmental reporting often starts with the selection of a few key indicators that have both ecological and management relevance (see Bockstaller and Girardin, 2003; e.g., Wascher, 2000; Lefebvre et al., 2005). In some cases that selection is harmonized with indicators of other ecosystems, such as urbanized or forested landscapes (e.g., McKenzie et al., 1992; National Research Council, 2000; Heinz Center, 2002). The critical challenge is to reconcile the scale at which ecosystem services are measured relative to agricultural management activities, which occur at the subfield to farm scale, and the effects of those activities, which are manifested at field, landscape or watershed, and regional scales.

The National Science Foundation recently (October 2005) organized a national conference on "Valuation of Ecosystem Services in Agriculture." A series of papers from this conference will be published in the journal *Ecological Economics*.

Landscape ecology

As land use intensity increases and becomes more pervasive, interactions between agricultural and natural ecosystems become more important, and landscape ecological approaches to monitoring and assessment may help decide what to measure and how. In spatial ecology, the "pattern-process" hypothesis (e.g., Turner, 1989; Turner et al., 2001) leads directly to monitoring agricultural landscapes with indicators of spatial pattern (Graham et al., 1991; O'Neill et al., 1997)—where the spatial pattern of landscape elements change, the ecological processes embedded in those landscapes can change accordingly. More generally, landscape ecology, as an interdisciplinary science (Naveh, 2001, Tress and Tress, 2001), incorporates hierarchical theory (Allen and Starr, 1982; O'Neill et al., 1986) and can provide a framework to harmonize indicator theory and application across the lines that divide agricultural from social and ecological sciences (e.g., Wu and Hobbs, 2002; Müller and Lenz, 2006).

Table 1. Selected examples of agricultural impacts and their resulting environmental concerns.

Concerns resulting from these impacts		Impacts from agriculture	Loss of habitat	Change in climate (from landscape modification)	Changes in hydrology	Changes in the ecosystem	Loss of soil (erosion)	Loss of soil fertility	Loss of water resources (subsurface storage)	Contamination	1. Chemical	a. nutrients	b. pesticides	c. antibiotics	d. trace elements	e. BOD	f. NOx and CO2	g. NH3 gas and air particles	2. Biological	a. microorganisms	b. GMO remnants	3. Physical	a. sediment	b. heat
Recreation	Fishing		X		X		X				X						X						X	X
	Swimming				X		X				X						X			X			X	
	Hunting		X		X								X											
	Hiking, ascetics		X	X		X												X						
Drinking water											X	X		X						X				
Human health												X		X						X	X			
Ecosystem health (toxicity, …)												X	X	X										X
Antibiotic resistance														X										
Ecosystem sustainability			X	X	X	X			X								X				X			
Agriculture sustainability				X			X	X	X															
Flooding			X		X																			
Water shortages					X				X															
River navigation							X																X	
Reservoir filling							X																X	
Gulf of Mexico hypoxia																								
Global climate change																	X							

Light gray shading = outcome due to legacy activities.

Medium gray shading = outcome partially due to legacy activities and partially due to ongoing activities.

Dark gray shading = outcome due to ongoing activities.

As a practical matter, the various branches of landscape ecology have not been integrated in an operational sense; as a result, landscape ecology provides neither a "methods manual" for monitoring, nor a "handbook" for agricultural land management. But landscape ecology can still supply conceptual approaches to the problem, and at least four key principles can be identified for monitoring, no matter what indicators or measurements are ultimately chosen:

1. *Spatial arrangement.* Real landscapes are neither perfectly regular nor completely ran-

dom, and the spatial arrangement of landscape elements regulates the flow of matter, energy, and information within and among landscapes. All observations must be mapped so that spatial variance can be explicitly analyzed and potentially related in space to other measures of ecosystem structure and function. For example, in agricultural landscapes, ideas about spatial arrangement have been extensively used in the design and implementation of riparian buffers to absorb pollutants moving from upland areas toward streams in surface and groundwater flow (Lowrance et al., 1997).

2. *Scale and hierarchy.* According to both general systems theory and hierarchy theory, it is convenient to organize an ecosystem analysis according to the characteristic spatial and temporal scales at which various parts of the ecosystem operate. Furthermore, a given question about ecosystem structure and function should be addressed at the most appropriate scale because it is rarely feasible to "scale up" from observations that are too detailed, and measurements at scales that are too coarse rarely provide enough information to guide management of specific tracts of land. Put more concretely, a scaled answer is probably wrong, and the only way to validate it is to measure the variable of interest at the large scale. On the other hand, mechanistic understanding of large-scale processes and the ability to affect change in those processes with management requires both up- and down-scaling. Landscape ecology provides a body of concepts, many adapted from general systems theory, hierarchical theory, and complexity theory, that provides guidance for these scaling activities.

3. *Context versus contents.* One of the most useful tools for scaling that comes from landscape ecology is the idea that a landscape can be described by the elements that it contains, or the same elements can be described by the landscapes that surround them. While the "contents approach" is more common, it should be avoided because it fixes the observation scale of the analytical unit (e.g., a watershed, county, or ecoregion), and the observations are difficult to extrapolate to other types of analytical units. This has been

termed the "modifiable area unit problem" in geography, whereby the aggregate answer depends upon how the pieces are aggregated. In contrast, the "context approach" does not pre-define assessment units and thus preserves options because the measurements can be aggregated in many different ways after measurements are taken. The context approach is more amenable to scaling because the results it produces are less site-specific.

4. *Landscape legacies.* Landscapes are like planets in the sense that each one has a unique historical trajectory. The future trajectories of two landscapes, however, can be different even if their current characteristics are similar. Individual landscapes that have become self-organized or managed around particular sets of governing processes can exhibit management responses that differ from other landscapes where different ecosystem drivers and constraints govern ecosystem functions. An example of an important legacy in agricultural landscapes is sediment that has accumulated in stream channels over many decades and contributes to contemporary impairment of water quality and delivery to receiving waters.

Adaptive management

Adaptive management is defined as "a process combining democratic principles, scientific analysis, education, and institutional learning to increase our understanding of ecosystem processes and the consequences of management interventions, and to improve the quality of data upon which decisions must be made" (Holling, 1978; Walters, 1986). It represents an iterative process in which monitoring and modeling are used to evaluate management activities, with consequent adjustments in models and management. Perhaps the greatest utility of adaptive management is that it acknowledges that no "cookbook" or specific "recipe" exists for "what to measure" in agricultural landscapes. Rather, "a thinking person's approach" is required to build a comprehensive understanding of a complex system, adjust models, and continually improve management.

Conceptual challenges

The collective actions of numerous land managers at the farm and field scale relate closely to ecosystem services, particularly water quality at the landscape scale. Because of the diffuse (non-point-source) nature of agricultural pollutants and historically "assigned" rights for farmers to pollute, agriculture has become a major source of water pollution and significantly degraded ecosystem services in many landscapes, leading to striking water quality crises in some agricultural regions. For example, in May 2000 several people died in the town of Walkerton, Ontario, Canada. The cause of the tragedy was identified as *E. coli* from agricultural sources that polluted the drinking water, and the public water testing system failed to detect the pollutant (de Loë and Kreutzwiser, 2005; Schuster et al., 2005). This incident led to creation of a source-water protection plan and a nutrient management act in Ontario that more stringently regulates farming practices.

On the one hand, public awareness of crises like what occurred in Walkerton has made farmers more liable for the degradation of water bodies and related ecosystem services. On the other hand, threats of agricultural pollution of water resources in rural communities also motivate some farmers or land managers to take responsibility and implement conservation practices. For example, the Otter Lake watershed in southern Illinois has suffered considerable water quality degradation since the early 1990s that has threatened the water supply and recreational resources of local communities. Monitoring data on atrazine and sediment revealed a strong relationship between specific farming activities and lake water quality. This information, coupled with risk assessment information, was provided to farming and nonfarming communities to help build consensus on cooperative conservation planning among stakeholders. Farmers then initiated conservation actions, such as reducing chemical use, implementing conservation tillage, and installing riparian buffers, voluntarily or with cost assistance, to improve water quality (Salamon et al., 1998). As a result, ecosystem services, including water quality in Otter Lake, improved considerably after a few years of conservation action.

These examples illustrate the value of considering diverse ecosystem services to generate consensus on solutions to problems with significant spatial- and temporal-scale complexities. The challenge is to develop assessment tools capable of determining if improvements in these services really are occurring in response to specific conservation actions.

A major challenge in applying ecosystem-service and landscape-ecology approaches to agricultural environmental performance issues is that individual farmers or land managers may not see the direct link between their operations and provision of ecosystem services for several reasons. Spatially, numerous farmers at various locations collectively affect ecosystem services. Therefore, the role of any individual is difficult to assess. Establishment of wildlife habitat, for example, may involve the efforts of many land managers to improve both the amount of green space and connectivity on an agricultural landscape. Temporally, ecosystem services, such as water quality improvement, may relate to conservation actions taken many years earlier. The scale of observation may further obscure the connection between land managers' actions and ecosystem services. For example, it would be difficult to make a connection between conservation practices at specific locations in the Mississippi River Basin with hypoxia reduction in the Gulf of Mexico (Ribaudo et al., 2001). From the farmer's perspective, it would be desirable to have a better understanding of these ecosystem complexities so that a link between an individual land manager's actions and specific ecosystem services could be established. Those tasks demand monitoring and modeling of both land management and associated environmental and ecosystem indicators.

Some ecosystem services from land stewardship may generate private benefits to farmers or land managers. But most ecosystem services produce societal benefits. For example, farmers who adopt conservation tillage, implement nutrient management, and establish riparian buffers may have higher soil quality, improved aesthetic values, and some economic benefits on their farms. But significant benefits, such as improved carbon sequestration, water quality, aquatic health, and wildlife habitat, are benefits to society at large. The asymmetry of benefit distribution, coupled with the private costs of conservation actions, has led to the establishment of public-funded conservation programs, such as the Conservation Reserve Program in the United States and the

Greencover Program in Canada. Those programs provide financial incentives to farmers for implementing conservation practices on their farms. Studies show that the societal benefits of land managers' conservation actions are far greater than the economic costs (Feather et al., 1999).

With the vastness of agricultural areas and limited public funds for conservation programs, a critical policy question confronts decision-makers (or regulators): How to allocate public funds selectively to achieve socially optimal benefits with a given budget? The cost effectiveness of existing conservation efforts also needs to be evaluated (Yang et al., 2003). These challenging tasks demand an understanding of the link between farm- and field-scale decisions and landscape-scale ecosystem services. For example, to achieve a specified level of water quality benefits in a watershed, policymakers need to identify pollution source areas and allocate public funds for farmers to implement conservation practices, such as no-till, nutrient management, and riparian buffers in targeted locations. If habitat improvement goals are also considered, practices to improve habitat amount and connectivity also need to be included in the design of conservation programs. Targeting and evaluating conservation programs at the landscape scale also require monitoring land management changes and related impacts on ecosystem services. In response to these needs, the Conservation Effects Assessment Project (CEAP) and the Watershed Evaluation of Beneficial Management Practices (WEBs) project have been established in the United States and Canada, respectively. Those projects are developing models (primarily hydrologic) that link monitoring results with landscape processes. The monitoring and modeling results are important for prioritizing locations in agricultural landscapes to implement conservation practices. This targeting is complicated by the biophysical complexity of watersheds and landscapes and also by the variation in land management practices, such that small areas of poorly managed land can account for a high percentage of the environmental quality problems in a watershed.

Scale and hierarchical concepts in landscape ecology are helpful for understanding the relationship between conservation programs and ecosystem services. While monitoring and modeling at the watershed or landscape level can provide information for targeted implementation of conservation practices, similar efforts at a regional scale (such as the Chesapeake Bay Basin and the Canadian prairies) also are of great importance for making conservation program decisions. Regional efforts may not be very detailed, but they do address a fundamental problem by identifying specifically which small watersheds or landscapes actually merit any effort at all.

For example, the Illinois Conservation Reserve Enhancement Program was established in 1998. With an approximate budget of $500 million, the program aims to retire 92,800 hectares (232,000 acres) of cropland within the 6.3-million-hectare (15.7-million-acre) Illinois River Basin to achieve multiple environmental and ecological benefits. The program objectives include reducing total sediment loading to the Illinois River by 20 percent; reducing phosphorus and nitrogen loading by 10 percent; increasing populations of waterfowl, shorebirds, and state and federally listed threatened and endangered species by 15 percent; and increasing native fish and mussel stocks by 10 percent (U.S. Department of Agriculture, 1998). This policy region (the Illinois River Basin) is composed of more than 100 U.S. Geological Survey 11-digit watersheds. Allocating public funds and conservation responsibilities (such as land retirement) across such a large area is a significant policy challenge, more difficult than conservation targeting within a particular watershed.

Targeting decisions at the regional scale must be based on robust monitoring and/or modeling at the regional scale, with mechanisms for downscaling information from the regional level to the local level to facilitate decision-making. Moreover, those decisions and the scaling and hierarchical challenges differ between the physical-chemical (sediment, phosphorus, nitrogen) and biological-ecological aspects (birds, fish) of the problem. Because the regional problems that arise from the physical and chemical impacts are the sum of inputs from each field or subfield, the inputs from the 11-digit watersheds can be summed to the whole Illinois River watershed for a regional-scale assessment. Working at a small scale to identify the major sources of sediment, nitrogen, and phosphorus is very important because most sediment and phosphorus probably comes from a small part of the landscape. In contrast, the wildlife issues require consideration of a completely different set of factors that operate at inherently larger scales, for example, connectivity between habitat patches and the availability of source populations for specific species. Clearly, diverse scales of investigation must be considered to encompass multiple ecosystem services within a particular region.

Environmental observations

Agriculture can be extremely varied (row crops, orchards, animals, forests, fish, etc.); therefore, agriculture's environmental impacts likewise are extremely varied. Table 1 lists some examples of agricultural impacts and their resulting environmental concerns. For a given landscape, usually only one or a few of those related impacts or concerns are of primary interest, but in reality, there usually are many impacts or concerns occurring at all times. The focus on only one or a few primary impacts or concerns is a natural result of limited resources for conservation and current public, political, and economic pressures. Changing knowledge (e.g., recognition of new pollution concerns) and changes in human values and preferences (e.g., new interests in carbon sequestration) also complicate focusing on monitoring efforts. Unfortunately, at times, numerous investigators and government agencies address different environmental concerns within a given landscape, but in relative isolation from each other. This can be attributed to research interests, agency missions, and lack of communication.

The sources of most environmental impacts are place-based at a small scale on the landscape, for example, within individual fields, orchards, and animal feeding operations, and at certain times across the agricultural cycle. Oftentimes, but not always, the resulting concerns also are place-based, but generally at a larger scale. As an example, soil erosion in a watershed oftentimes impacts streams with elevated sediment and phosphorous concentrations. A majority of this erosion comes from a small percentage of the watershed (generally areas of substantial topography near the stream network), but is perceived as a watershed-wide concern. The development of a plan for making environmental observations (including routine monitoring and targeted sampling) must consider the pertinent temporal and spatial scales for the source, transport, and receiving portions

of the environment. Sampling strategies that leave out one or the other of these factors may provide inadequate information for making good management decisions. Also, sampling strategies that do not consider the important differences in various portions of the landscape, by working at too large of a scale or by considering only average characteristic values (such as slope or soil type), may have the same problems.

In the social dimension, there is an ever-growing need for better time-varying, placed-based information on such variables as cropping patterns, animal operations, agricultural management practices, chemical use, and so forth to help make better decisions. Gathering data on these social variables raises concerns about protection of privacy.

Development of a plan for making environmental observations can be based on an iterative series of questions. The most important question is "why." That is: "What is the purpose of the data collection plan?" Common purposes include the following: To make an assessment of the environment, to quantify the results of a management action, to fulfill a regulatory requirement, to estimate loads or trends, to understand environmental processes, or to use the measured parameter as a surrogate for parameters that are difficult to measure. The purpose should then drive the rest of the sampling plan. Other important questions include:

- What should be measured, given the possible range of physical, chemical, biological, hydrological, societal, and other parameters?
- Which environmental compartment(s) should be included, such as soil, surface waters, groundwater, atmosphere, biota, and people?
- When should sampling take place, given diurnal and seasonal considerations and the frequency of data needed?
- Where specifically should measurements be made that will provide data representative of the landscape of interest?
- How should measurements be made, including field methods, laboratory methods, and quality control considerations?

Given the numerous ways in which this series of questions can be answered, there are no set formulas that can be applied directly to making environmental observations for a specific set of concerns in a given landscape. The scientific literature is full of examples that help answer these questions. There also are a number of large-scale studies that include agricultural landscapes and concerns that demonstrate that high quality environmental observations can be made in a consistent manner over large spatial and temporal timeframes. These studies include the following: The U.S. Environmental Protection Agency Environmental Monitoring and Assessment Program (EMAP), the U.S. Geological Survey's National Water-Quality Assessment (NAWQA) program, the U.S. Geological Survey's Biomonitoring of Environmental Status and Trends (BEST) program, the U.S. Department of Agriculture's Conservation Effects Assessment Program (CEAP), and Canada's National Agri-Environmental Health Analysis and Reporting Program (NAHARP). Some of these studies and others (for example, U.S. Department of Agriculture-Natural Resources Conservation Service, Heinz Center, and National Research Council Commission on Geosciences, Environment and Resources) have developed indices that guide the choice of and help interpret environmental observations to make informed management decisions about the environment.

Ecosystem services provide an approach for answering the "why" "what" and "which" questions above. If the list and rank of the services of interest can be articulated, the parameters and compartments to be measured will be known.

Landscape ecology offers guidance to help answer the "when" and "where" questions. At a basic level, landscape ecology suggests that if the "where" and "when" questions are important the question is being asked at the wrong scale. For example, if annual delivery of nitrogen to the Gulf of Mexico is the concern, the annual delivery of nitrogen to the Gulf of Mexico should be measured, rather than trying to multiply values measured in individual farm fields by some enormous number. On the other hand, if the objectives are to understand the factors controlling this delivery and design management programs to reduce it, robust scaling methods (based on concepts from landscape ecology) are needed to design sampling schemes that account for field-by-field dynamics. For example, chemical and hydrologic questions require consideration of at least three time frames (the "when" question): Seasonal (temperature, portion of growing season, times of chemical

applications, and field activities), hydrologic stage (high flow, low flow), and diurnal cycles (especially for dissolved oxygen and other biologically active parameters). While landscape ecology can help to address these time frames in a general sense, process-level, mechanistic understanding also is required to design adequate sampling schemes.

It is well understood that all aspects of the environment, including agricultural activities, are interconnected. The hydrologic system and the interaction of the atmosphere, terrestrial, and aquatic environments frequently are the important links in these cause-effect relationships. An examination of Table 1 suggests that almost all concerns relate directly or indirectly to water. Because this linkage is so strong, it becomes imperative that landscape hydrology be understood to make informed decisions on how best to manage the landscape for environmental quality. Understanding hydrology involves understanding the environmental pathways, fluxes, and residence times of water within the landscape and exchanges that occur within adjacent areas, regional aquifers, and the atmosphere. For example, compare nitrate in streams that have different hydrologic settings. One landscape is characterized by fine soils and a deep aquifer that is not hydrologically connected to the stream. The other landscape is characterized by permeable soils underlain by a sand aquifer that is connected to the stream. The most probable route of nitrate transport from the field of the first stream is direct runoff to the stream. In the other case, the most probably route is leaching to the water table, transport through groundwater, and eventual movement into the stream. Understanding the differences in these two hydrologic settings will suggest preferred management practices (strategies that control runoff will be effective for the former, but counterproductive for the later) and provide realistic expectations of the solution. The lag time between implementation of the solution for the runoff-dominated landscape and the observed reduction in nitrate concentration in the stream should only be until the next runoff event. In the groundwater-dominated landscape, the lag time could be years to decades, depending upon residence time in the aquifer, before a change in nitrate concentration might be observed.

Superimposed on hydrology are a myriad of

physical, chemical, and biological processes and interactions that must be understood to make informed environmental management decisions. Many of these processes have been studied in the laboratory and/or field. Results from those studies form the basis for a general understanding that, hopefully, can be transferred to the landscape of interest. Assessment of the adequacy of this transference oftentimes will drive decisions about environmental observations. For example, the process of denitrification is well understood as a strong function of environmental redox conditions. Therefore, an aquifer that has a nitrate concern should be monitored for nitrate, but also redox indicators, such as dissolved oxygen and dissolved iron. In addition, an estimate of travel and residence time in the aquifer would provide a context for assessing the potential impacts of the nitrate on the receiving surface water body.

Modeling

Although environmental observations provide the foundation for assessing and understanding the impacts of agricultural activities on the landscape, there are practical spatial and temporal limitations to the extent of field observations. In cases where it is not feasible to make direct field observations, mathematical simulation models can be used to supplement and extend (but not replace) field observations. Models can be used to aid interpretation and extrapolation of field observations and/or make predictions for unstudied areas. In most cases, a number of different models are needed to address adequately the diverse range of environments and environmental concerns.

Getting the water right

If the idea is accepted that hydrology is central to many, if not most, agricultural and environmental interactions, we first must ask if our models can "get the water right." Models must depict both surface (e.g., runoff) and groundwater recharge processes, something that is rare in a single model. A second challenge is scaling from fields to watersheds to regions. While there are several models capable of integrating across field and landscape or watershed scales (Arnold et al., 1998; Renschler and Lee, 2005), encompassing interaction between landscape elements is an

even greater challenge. Interaction between elements is addressed in some specialized models [e.g., the Riparian Ecosystem Management Model (REMM)], (Lowrance et al., 2000) and in "eco-hydrological" models that are spatially explicit depictions of the landscape, driven by detailed terrain models (Running et al., 1989; Band et al., 1993; Beven and Kirkby, 1997). A final hydrologic challenge is to encompass "in-stream" processes, including physical redistribution of sediments and redistribution of the stream channel (Simon, 1989; Langendoen, 2000), as well as biotic-nutrient processing (Peterson et al., 2001).

Great progress is being made in hydrologic modeling of the environment under the aegis of the WATERS Network (an initiative in the National Science Foundation's Engineering and Geosciences Directorates; http://www.cuahsi.org and http://cleaner.ncsa.uiuc.edu/). That network seeks to establish "hydrologic observatories" to improve the predictive understanding of flow paths, fluxes, and residence times of water, sediments, and nutrients across a range of spatial and temporal scales. Those observatories will lead to creation of "digital watersheds," an assembly and synthesis for a hydrologic region of point hydrologic observations and geographic information system, remote sensing, and weather- and climate-gridded data. The intent is to have a comprehensive digital description of the physical environment and water conditions within a hydrologic observatory. Those observatories have the potential to aid greatly the analysis of agricultural and environmental interactions.

Once we are convinced that our models can "get the water right," the next challenge is to interface hydrologic models with models capable of depicting biologically based ecosystem services. Ecohydrologic models link hydrology with biogeochemical processes, transforming nutrients as they move across the landscape with water (Band et al., 1993). These models show great promise for depicting processes such as denitrification, especially if they are linked to data streams from hydrologic observatories and digital watersheds.

A greater challenge is to establish links between hydrology and other biological variables of interest in an ecosystem services context, for example, wildlife habitat (Santelmann et al., 2006). Biologically based goals and services are not articulated as well as goals for hydrology and chemistry. The processes are not as well understood, and modeling and monitoring protocols are not as well developed. Current proposals to develop a National Ecological Observatory Network (NEON, www.neoninc.org) have the potential to transform our ability to understand and predict biologically based ecosystem services. NEON is envisioned as a continental-scale research instrument consisting of geographically distributed infrastructure, networked via state-of-the-art communications. If implemented, it will provide critical environmental measurements, experiments, and models to improve understanding and predictability of key ecological processes and services.

Including people

Ultimately, models must incorporate social science variables that influence adoption of best management practices (Wagner, 2005) that produce output on the broad range of ecosystem services relevant to diverse groups of stakeholders (El-Swaify and Yakowitz, 1998; Costanza et al., 2002; Lant et al., 2005) and link to decision-support and policy planning tools. Current models have limited capacity to provide these outputs.

Improvements in social science models require better acquisition of time-varying, placed-based information on cropping patterns, animal operations, agricultural management practices, chemical use, and so forth on multiple scales. More fundamentally, understanding of the significant variation in human management must be improved, even with the same location and type of cropping system. Improvement in this area clearly requires new conceptual approaches as well as new vehicles for collecting and analyzing data, for example, remote sensing, microsensors, and automated reporting systems.

The importance of social science as a critical component of integrated environmental analysis has been recognized for the past 10 to 20 years (Groffman and Likens, 1994). It is interesting to note that the National Science Foundation's Long-term Ecological Research network is embarking on a planning effort that focuses on the links between ecosystem structure and function (the traditional purview of the network) and ecosystem services and, more importantly, on human perceptions of those services and how those per-

ceptions alter human behavior that influences ecosystem structure and function. Application of a similar framework to agricultural landscapes would likely be quite useful.

Model quality

Linking a large number of models from diverse fields creates great challenges for integration and validation. New approaches to "modular modeling" (Voinov et al., 2004), such as the advanced Modular Modeling Framework for Agricultural and Natural Resource Systems developed during a 1997 interagency workshop (Agricultural Research Service, Natural Resources Conservation Service, and U.S. Geological Survey) and those being developed by the Interagency Steering Committee on Multimedia Environmental Models (www.iscmem.org), should yield progress over the next few years. Development of standards for validation of multidisciplinary modular models, similar to those developed by the American Society for Testing and Materials (ASTM, 1984), for environmental fate models will be particularly critical if those models are going to be used for management and/or regulatory assessments. The ASTM evaluation standards have five components: Model examination, algorithm examination, data evaluation, sensitivity analysis, and validation.

An example: The Chesapeake Bay Program

Perhaps the best example of an integrated, iterative, goal-driven adaptive management-based program of environmental monitoring and modeling is the Chesapeake Bay Program (Boesch et al., 2001, www.chesapeakebay.net). In 1983, a multistate agreement set specific goals for reducing multiple sources of nutrients to the bay. Progress toward reaching those goals has been carefully tracked by an extensive monitoring plan, with a comprehensive model used to guide the monitoring and synthesize results. Over time, goals have been reevaluated, the model and monitoring programs have been adjusted, and critical uncertainties have been identified. Identifying critical uncertainties has helped to articulate a diverse set of research needs ranging from field-scale best management practices to ecosystem-service driven ecological economic modeling

(Costanza et al., 2002).

The ongoing effort in the Chesapeake Bay Basin illustrates the challenge that we face in managing agricultural and environmental interactions. Despite more than 20 years of high-profile (and expensive) effort, many goals for nutrient reductions and improvement of living resources and ecosystem services have not been met, and the overarching model and specific programs have been criticized, reaching the level of a congressional hearing (Whoriskey, 2004). But the controversy surrounding the Chesapeake Bay Program may be a testament to its inherently robust approach, where monitoring data are used to evaluate progress in achieving goals and to motivate new research and reevaluation of goals and models. It is a dynamic, evolutionary, and iterative adaptive management process that is directly addressing a complex and important problem.

Summary and conclusions

This is an exciting time in agricultural and environmental science. New ways of looking at landscapes and the multiple ecosystem services they produce and new tools for compiling, integrating, and modeling environmental data lead us to conclude that landscape-scale assessment of multiple ecosystem services can become "normal operating procedure" for conservation agencies within the next 10 to 20 years. To be effective, this procedure must involve an iterative adaptive management process driven by a clear articulation of goals for comprehensive sets of ecosystem services at the farm, landscape, watershed, and regional scales, coupled to monitoring and modeling programs that evaluate how well those goals are being attained, and followed by a reevaluation of goals and management strategies and continued monitoring and modeling. A dynamic, evolutionary process that links monitoring, modeling, and research is the only way that the complex nature of agricultural and environmental interactions can be addressed that lead to real improvements in environmental quality on agricultural landscapes.

Several key research needs must be addressed to facilitate development of multiobjective, multiscale assessment procedures. The science behind ecosystem services, in particular quantification and valuation of "cultural" and "supporting" services, needs further conceptual and practical development. Relationships between conservation programs and ecosystem services are complex. Uncertainties caused by such factors as climate and on- and off-site characteristics and their interactions mandate a careful, cautious approach by land managers and regulators. Understanding the nature and extent of variation in human behavior is perhaps the greatest research challenge, for example, we need to know where, why, and how a small percentage of land managers produce a significant portion of the environmental problems in a given area.

Another major question is this: If both the complicated aspect and complexity of agricultural and environmental interactions increase in a nonlinear fashion as we move from the field to the landscape to the regional scale, will we be more effective and efficient as both scientists and policymakers if we work at the regional scale without trying to understand the underlying nonlinearities, or do we need to take a reductionist approach and focus on small-scale watersheds? Or should we be prepared to do both, depending upon the specific concern of interest?

For any given concern, we need to determine the preferred scale(s) for monitoring and planning to enhance ecological goods and services currently being degraded from agricultural production in an efficient and effective way. Barring some great conceptual advance in our understanding of scaling, we suggest that monitoring and modeling programs must encompass small (subfield) to landscape and regional scale dynamics. Such a broad-scale effort will need to be facilitated by improvements in monitoring technology (e.g., wireless sensors, remote sensing) and information technology (cyberinfrastructure). Our role as scientists will be to articulate just where these improvements are needed and how they can be incorporated into monitoring programs linked to comprehensive models. Efforts such as WATERS and NEON have great potential to address these key monitoring and modeling challenges and should lead to real progress over the next decade or two.

References

Allen, T.F.H., and T.B. Starr. 1982. Hierarchy: Perspective for ecological complexity. University of Chicago Press, Chicago, Illinois.

Arnold, J.G., R. Srinivasan, R.S. Muttiah, and J.R. Williams. 1998. Large area hydrologic modeling and assessment, Part I: Model development. *Journal of the American Water Resources Association* 34(1):73-89.

ASTM. 1984. Standard practice for evaluating environmental fate models of chemicals. Standard E 978-84. American Society for Testing and Materials, Philadelphia, Pennsylvania.

Band, L.E., J.P. Patterson, R. Nemani and S.W. Running. 1993. Forest ecosystem processes at the watershed scale: Incorporating hillslope hydrology. *Agricultural and Forest Meteorology* 63:93-126.

Beven, K.J., and M.J. Kirkby. 1997. A physically-based variable contributing area model of basin hydrology. *Hydrological Sciences Bulletin* 24:4369.

Bockstaller, C., and P. Girardin. 2003. How to validate environmental indicators. *Agricultural Systems* 76:639-653.

Boesch, D.F., R. Brinsfield, and R. Magnien. 2001. Chesapeake Bay eutrophication: Scientific understanding, ecosystem restoration, and chellenges fo ragricutlure. *Journal of Environmental Quality* 30: 303-320.

Bouma, J. 2002. Land quality indicators of sustainable land management across scales. *Agriculture, Ecosystems & Environment* 88:129-136.

Clergue, B., B. Amiaud, F. Pervanchon, F. Lasserre-Joulin, and S. Plantureux. 2005. Biodiversity: function and assessment in agricultural areas, a review. *Agronomy for Sustainable Development* 25:1-15.

Carpenter, S.R., and M. Turner. 2000. Opening the black boxes: Ecosystem system and economic valuation. *Ecosystems* 3:1-3.

Costanza, R., R. d'Arge, R. De Groot, S. Farver, M. Grasso, B. Hannon, K. Limburg, S. Naeem, R.V. O'Neill, J. Paruelo, R.G. Raskin, P. Sutton, and M. van den Belt. 1997. The values of the world's ecosystem services and natural capital. *Nature* 387:253-260.

Costanza, R., A. Voinov, R. Boumans, T. Maxwell, F. Villa, L. Wainger, and H. Voinov. 2002. Integrated ecological economic modeling of the Patuxent River watershed. *Ecological Monographs* 72: 203-231.

Daily, G.C., S. Alexander, P.R. Ehrlich, L. Goulder, J. Lubchenco, P.A. Matson, H.A. Mooney, S. Postel, S.H. Schneider, D.G. Tilman, and G.M. Woodwell. 1997. Ecosystem services: benefits supplied to human societies by natural ecosystems. *Issues in Ecology* 2:1-16.

DeFries, R.S., J.A. Foley, and G.P. Asner. 2004. Land-use choices: Balancing human needs and ecosystem function. *Frontiers in Ecology and Environment* 2:249-257.

de Loë, R.C., and R.D. Kreutzwiser. 2005. Closing the groundwater protection implementation gap. *Geoforum* 36:241-256.

El-Swaify, S.A., and D.S. Yakowitz. 1998. Multiple objective decision making for land, water, and environmental management. Lewis Publishers, Boca Raton, Florida. 743 pp.

Feather, P., D. Hellerstein, and L. Hansen. 1999. Economic evaluation of environmental benefits and the targeting of conservation programs: The case of the CRP. Agricultural Economic Report No. 778. Economic Research Service, U.S. Department of Agriculture, Washington, D.C.

Freyenberger, S., R. Janke, and D. Norman. 1997. Indicators of sustainability in whole-farm planning: Literature review. Kansas Sustainable Agriculture Series, Paper #2, Contribution No. 97-482-D. Kansas Agricultural Experiment Station, Kansas State University, Manhattan.

Graham, R.L., C.T. Hunsaker, R.V. O'Neill, and B.L. Jackson. 1991. Ecological risk assessment at the regional scale. *Ecological Applications* 1:196-206.

Groffman, P.M., and G.E. Likens, editors. 1994. Integrated regional models: Interactions between humans and their environment. Chapman and Hall, New York, New York.

Groffman, P.M., C.T. Driscoll, G.E. Likens, T.J. Fahey, R.T. Holmes, C. Eagar, and J.D. Aber. 2004. Nor gloom of night: A new conceptual model for the Hubbard Brook ecosystem study. *BioScience* 54:139-148.

H. John Heinz III Center for Science, Economics and the Environment. 2002. The state of the nation's ecosystems: Measuring the lands, waters, and living resources of the United States. Cambridge University Press, Cambridge, United Kingdom.

Holling C.S., editor. 1978. Adaptive environmental assessment and management. John Wiley & Sons, London, England.

Langendoen, E.J. 2000. CONCEPTS—Conservational Channel Evolution and Pollutant Transport System. Research Report No. 16. National Sedimentation Laboratory, Agricultural Research Service, U.S. Department of Agriculture, Oxford, Mississippi. 160 pp.

Lant, C.L., S.E. Kraft, J. Beaulieu, D. Bennett, T. Loftus, and J. Nicklow. 2005. Using GIS-based ecological-economic modeling to evaluate policies affecting agricultural watersheds. *Ecological Economics* 55:467-484.

Lefebvre, A., W. Eilers, and B. Chunn, editors. 2005. Environmental sustainability of Canadian agriculture: Agri-environmental indicator report series – report #2. Agriculture and Agri-Food Canada, Ottawa, Ontario.

Lowrance, R., L.S. Altier, J.D. Newbold, R.R. Schnabel, P.M. Groffman, J.M. Denver, D.L. Correll, J.W. Gilliam, J.L. Robinson, R.B. Brinsfield, K.W. Staver, W. Lucas, and A.H. Todd. 1997. Water quality functions of riparian forest buffers in Chesapeake Bay watersheds. *Environmental Management* 21:687-712.

Lowrance, R.R., L.S. Altier, R.G. Williams, S.P. Inamdar, D.D. Bosch, R.K. Hubbard, and D.L. Thomas. 2000. The riparian ecosystem management model. *Journal of Soil and Water Conservation* 55:27-36.

McKenzie, D. H., D. E. Hyatt, and V. J. McDonald, editors. 1992. Ecological indicators. Elsevier, New York, New York.

Müller, F., and R. Lenz. 2006. Ecological indicators: theoretical fundamentals of consistent applications in environmental management. *Ecological Indicators* 6:1-5.

National Research Council. 2000. Ecological indicators for the nation. National Academies Press, Washington, D.C.

Naveh, Z. 2001. Ten major premises for a holistic conception of multifunctional landscapes. *Landscape and Urban Planning* 57:269-284.

O'Neill, R.V., D.L. De Angelis, J.B. Waide, and T.F.H. Allen. 1986. A hierarchical concept of ecosystems. Princeton University Press, Princeton, New Jersey.

O'Neill, R.V., C.T. Hunsaker, K.B. Jones, K.H. Riitters, J.D. Wickham, P. Schwarz, I. Goodman, B.L. Jackson, and W. Baillargeon. 1997. Monitoring environmental quality at the landscape scale. *BioScience* 47:513-519.

Peterson, B. J., W. M. Wollheim, P. J. Mulholland, J. R. Webster, J. L. Meyer, J. L. Tank, E. Marti, W. B. Bowden, H. M. Valett, A. E. Hershey, W. H. McDowell, W. K. Dodds, S. K. Hamilton, S. Gregory, and D. D. Morrall. 2001. Control of nitrogen export from watersheds by headwater streams. *Science* 292:86-90.

Renschler, C.S., and T. Lee. 2005. Spatially distributed assessment of short- and long-term impacts of multiple best management practices in agricultural watersheds. *Journal of Soil and Water Conservation* 60:446-456.

Ribaudo, M., R. Heimlich, R. Claassen, and M. Peters. 2001. Least-cost management of nonpoint source pollution: Source reduction vs. interception strategies for controlling nitrogen loss in the Mississippi Basin. *Ecological Economics* 37: 183-197.

Rodriguez, J.P., T. D. Beard, Jr., E.M. Bennett, G.S. Cumming, S.J. Cork, J. Agard, A.P. Dobson, and G.D. Peterson. 2006. Trade-offs across space, time, and ecosystem services. *Ecology and Society* 11:28.

Running, S.W., R.R. Nemani, D.L. Peterson, L.E. Band, D.F. Potts, and L.L. Pierce. 1989. Mapping regional forest evapotranspiration and photosynthesis by coupling satellite data with ecosystems simulation. *Ecology* 70:1090-1101.

Salamon, S., R.L. Farnsworth, and J.A. Rendziak. 1998. Is locally led conservation planning working? A farm town case study. *Rural Sociology* 63: 214-234.

Santelmann, M., K. Freemark, J. Sifneos, and D. White. 2006. Assessing effects of alternative agricultural practices on wildlife habitat in Iowa, USA. *Agriculture, Ecosystems and Environment* 113:243-253.

Schröder, J.J., D. Scholefield, F. Cabral, and G. Hofman. 2004. The effects of nutrient losses from agriculture on ground and surface water quality: the position of science in developing indicators for regulation. *Environmental Science & Policy* 7:15-23.

Schuster, C.J., A.G. Ellis, W.J. Robertson, D.E. Charron, J.J. Aramini, B.J. Marshall, and D.T. Medeiros. 2005. Infectious disease outbreaks related to drinking water in Canada, 1974-2001. *Canadian Journal of Public Health* 96: 254-258.

Simon, A. 1989. A model of channel response in distributed alluvial channels. *Earth Surface Processes and Landforms* 14:11-26.

Tress, B., and G. Tress. 2001. Capitalizing on multiplicity: a transdisciplinary systems approach to landscape research. *Landscape and Urban Planning* 57:143-157.

Turner, M.G. 1989. Landscape ecology: The effect of pattern on process. *Annual Review of Ecological Systems* 20:171-197.

Turner, M.G., R.H. Gardner, and R.V. O'Neill. 2001. Landscape ecology in theory and practice: pattern and process. Springer-Verlag, New York, New York.

van der Werf, H.M.G., and J. Petit. 2002. Evaluation of the environmental impact of agriculture at the farm level: a comparison and analysis of 12 indicator-based methods. *Agriculture, Ecosystems & Environment* 93:131-145.

Voinov, A., C. Fitz, R. Boumans, and R. Costanza. 2004. Modular ecosystem modeling. *Environmental Modelling and Software* 19: 285-304.

Wagner, M.M. 2005. Watershed-scale social assessment. *Journal of Soil and Water Conservation* 60:177-186.

Walters C.J. 1986. Adaptive management of renewable resources. McGraw Hill, New York, New York.

Wascher, D.M., editor. 2000. Agri-environmental indicators for sustainable agriculture in Europe. ECNC technical report series. European Centre for Nature Conservation, Tilburn.

Whoriskey, P. 2004. Progress minimal on bay pollution. The Washington Post, August 21, 2004, pp. B01.

Wu, J., and R. Hobbs. 2002. Key issues and research priorities in landscape ecology: An idiosyncratic synthesis. *Landscape Ecology* 17: 355–365.

Yang, W., M. Khanna, R. Farnsworth, and H. Onal. 2003. Integrating economic, environmental and GIS modeling to determine cost effective land retirement in multiple watersheds. *Ecological Economics* 46: 249-267.

Roundtable:
What should we measure, and how, to account for environmental effects?

Five themes emerged during the initial discussion at this roundtable:

1. Ensure that when the term "model" is used people are clear what is being talked about. Models can be simple or complex representations of system interactions. What is needed for what purpose?

2. The dilemma of a model being desired by the research community and real-life examples being desired by policymakers.

3. How are environmental benefits determined when surrogate indicators are measured?

4. Are data available to show where practices are installed? Isn't this information necessary to show "real" effects in real watersheds?

5. Can a biotic response be related to the location of a conservation practice?

Roundtable participants then identified six "next steps" important to strengthening the science of what to measure, and how, to account for environmental effects:

1. Gather success stories based on real data with a socioeconomic component.

2. Determine or quantify the economic value of outcomes at the local level associated with treating the land at the farm-community level.

3. Evaluate whether the phrase "ecosystem services" resonates with the public and policymakers before scientists and others make substantial investments in this concept.

4. Work with marketers to determine "if" and "how" icons might be useful in motivating people to act and/or change behavior?

5. Assist scientists and conservationists in developing better messages—ones that are simple, to the point, and resonate. The public and policymakers do not understand and are not interested in detailed, complex descriptions of models or research protocols.

6. Educate policymakers on the concept of "directionally correct investment and resulting accomplishments." Let scientists "fine tune" the decimal point, but ensure that conservation work is moving in the right direction.

Integrating the biophysical and social sciences

K.M. Sylvester
C.L. Redman

It is no longer tenable to study ecological and social systems in isolation of one another (Redman, 1999; Kinzig et al., 2001; Gunderson and Holling, 2002). Humans are an integral part of virtually all ecosystems (McDonnell and Pickett, 1993; Vitousek et al., 1997); almost all human activity has potential relevance to global environments (National Research Council, 1999); and biogeophysical contexts strongly condition human decisions (Diamond, 1997). Recently, the National Science Foundation underscored this when it released a 20-year review of the Long-Term Ecological Research (LTER) program. One of the main charges to the network was a recommendation for more collaboration with social scientists to increase the understanding of the reciprocal impacts of natural ecosystems and human systems. The review concluded that this was an essential next step in order to better inform environmental policy (National Science Foundation, 2002).

Agronomy, hydrology, and soil science have investigated human impacts, management, and cultural methods for much longer than ecology. But as ecology begins to play catch up, the discipline not only shares the agricultural sciences emphasis on place-based research and the analy-

sis of complex systems, but adds an important dimension. By paying attention to long cycles of change, there is much more emphasis in ecology on structural change over time and space, types and rates of change, scales of phenomena, strengths of linkages, boundary conditions, and threshold values (Carpenter, 1999). In short, it is the adaptive dimension of human and natural systems that ecology seems well-positioned to confront, even if the integrative frameworks are short on specifics. But the implications of thinking in terms of generational change or life-cycle change are important for monitoring and managing agricultural landscapes.

Monitoring

Let us briefly review the program of monitoring put forward in the foregoing section of this book by Peter Groffman and colleagues. Those authors make the case that monitoring is a necessary first step to convince policymakers and farmers that sustainability is not illusory—that it can be documented. They offer concepts from ecosystem services and landscape ecology as a framework from which to build monitoring tools—an iterative process of goal-setting, monitoring, modelling, assessment, and re-evaluation. To tell the story to the public, there must be benchmarks.

The first organizing idea for this monitoring framework is that it must pay attention to spatial location because the physical structure of landscapes regulates ecosystem process. Variance will be strongly related to location. Few practitioners in the environmental sciences would dispute the need for a place-based approach.

But they raise the stakes with the second idea that a monitoring system needs to be organized hierarchically. More than simply monitoring ecosystem processes at scales considered most appropriate—the field, the farm, the watershed, the landscape, the region—an effective framework will monitor the processes of interest at several scales simultaneously.

The third organizing idea is subtle and less obvious, but no less important. Making a distinction between the content and context of landscapes, the authors recommend that land managers remain flexible about the boundaries of systems. It is a different way of expressing the need for hierchical monitoring. The boundaries of systems need to be elastic in order to monitor trophic structure. When rapidly changing processes—nutrient cycles, organic matter accumulation, patterns of disturbance—alter landscape-scale processes, the evidence is not often visible at the field level. Monitoring needs coarse scales of observation to know when spillovers begin to affect landscape- or regional-scale structures. And those studies need time depth to understand the longevity of processes and their tipping points.

The fourth organizing idea is that historical trajectories are important. But here it is worth cautioning against frameworks that tend to treat trajectories as preordained. The language used to describe the effect of past land use suggests that trajectories are based on self-organized landscapes and governing processes. This language is probably too deterministic to begin the process of convincing policymakers and farmers that there are genuine alternatives to current forms of organization and that a program of social and ecological monitoring will help to uncover the scope for change. Yes, the room for variation is strongly constrained by environmental context. But self-organization implies a limited range of outcomes. It also takes us away from the surprises of history, the paths not taken but possible, the unanticipated collapses, or just from the tremendous complexity that led to the current dynamics of environmental change.

Scale

Part way into the paper in a section on the challenges of ecosystem services to watershed analysis, Groffman and his coauthors suggest that the private benefits of better land stewardship are often much smaller than the public benefits. The returns that a farm might see from conservation tillage or riparian buffers are not as great as the wider benefits to society of water quality and carbon sequestration. The asymmetry of benefit helps to explain why conservation programs like the Conservation Reserve Program in the United States and the Greencover Program in Canada are necessary for implementation to occur.

Yet, the authors subsequently acknowledge the vastness of the problem. Two-thirds of the earth's terrestrial surface is in agricultural land use. Fifty percent of the coterminous United States is cropped or grazed. How can public remediation

schemes possibly make a dent in the implementation of better management practices when the funds available for those programs are limited? Public schemes are important policy tools and especially useful in targeting sensitive ecosystem processes and locations. But in seeking broad implementation, it is essential that farmers themselves see the connections between practices and broader outcomes. Conservation science should not underestimate the potential for farm management to keep an eye on the long term. Farms are succession-minded institutions. To the extent that incentives for better management enhance the longevity of farm enterprises, they enhance the conservation of natural resources for future generations.

Social and ecological systems

Groffman and his colleagues take a somewhat pessimistic view of farm management. They are not specific about the social science variables needed to monitor the adoption of best management practices. So let us suggest some ways to conceptualize the interaction between social and ecological systems.

The current national-scale monitoring effort by the U.S. Department of Agriculture, for instance, will produce a benchmark study with unprecedented geographic breadth. But the Conservation Effects Assessment Project (CEAP) (Makuch et al., 2004) will not monitor social conditions that might be relevant to land use decisions. Nor are any return visits to the 30,000 cropping sample points [fields or land segments that vary in size from 16 to 256 hectares (40 to 640 acres)] currently planned. Data will be gathered on more than 200 attributes, including land use and land cover, soil type, cropping history, conservation practices, soil erosion potential, water and wind erosion estimates, wetlands, wildlife habitat, vegetative cover, and irrigation methods, and the four waves of data collection in 2003, 2004, 2005, and 2006 will produce a pooled data set that will drive hydrological models.

CEAP is an important study that will provide exhaustive background information. By not monitoring social change and not pursuing follow-up farm surveys, however, the benchmarking exercise may foreclose discussion about trajectories or the complexity of coupled human and natural systems. Ultimately, we want to be able to measure growth or decline and apportion variance to the scales at which change happens; that will allow us to see how much context shapes outcomes and how much room exists for adaptation and change.

There are many examples of longitudinal, multilevel, or agent-based analysis in the social sciences. For example, one of the longest running repeat surveys in the United States, the Panel Study on Income Dynamics, has been conducted at the University of Michigan since 1968. The original design consisted of two independent samples: A cross-sectional national sample (3,000) and a national sample of low-income families (2,000). From 1968 to 1996, the study interviewed and re-interviewed individuals from the families in the core sample every year. The study collected data on family composition changes, housing and food expenditures, marriage and fertility histories, employment, income, time spent in housework, health, consumption, wealth and more. The core sample grew to 8,500 families in 1996. One kind of analysis the data allow for is the intergenerational transfer of earning status. Think of what such a design could tell us about farm practices and landscapes. Of course, respondent identities would have to be protected and the landscape information presented in a nonidentifying way. But if nested in an informed way within a larger cross-sectional database that posed similar questions about farm practice, we would have the basic data needed to model how and when microlevel processes begin to affect macro-level structures and trends.

Integrated research

What are the aspects of human change that we need to monitor? Redman is a participant in the deliberations of the Resilience Alliance, which has established some first principles for researching integrated social-ecological systems and identifying adaptive capacity in them (www.resalliance.org). The alliance has proposed an integrated framework that parses natural and human spheres, but conceptualizes those spheres as components of a single, complex social-ecological system (Levin, 1999; Gunderson and Holling, 2002). The ecological patterns and processes are all things familiar to those involved with the LTER

network: Primary production, trophic structure, organic matter accumulation, inorganic nutrient flow, and disturbance. In the social realm, the proposed patterns and processes also focus on long-term dynamics. Cultural phenomena, framed by economic incentives, play a large role. Those cultural phenomena include, but are not limited to:

- *Demography*: The growth, size, composition, distribution, and movement of human populations.
- *Technological change*: The accumulated store of cultural knowledge about how to adapt to, use, and act upon the biophysical environment and its material resources to satisfy human needs and wants.
- *Economic growth*: The sets of institutional arrangements through which goods and services are produced and distributed.
- *Political and social institutions*: The enduring sets of ideas of how to accomplish goals recognized as important in a society. Family, religious, economic, educational, health, and political institutions that characterize its way of life.
- *Culture*: Culturally determined attitudes, beliefs, and values that purport to characterize aspects of collective reality, sentiments, and preferences of various groups at different scales, times, and places.
- *Knowledge and information exchange*: The genetic and cultural communication of instructions, data, ideas, and so on.

Many of these phenomena are straight out of the National Research Council's 1992 report on Global Environmental Change (1992:2-3), with one important exception—knowledge and information exchange (Berkes and Folke, 1998). Institutions, culture, and knowledge are difficult for physical scientists to incorporate into their research, but they are vital. All choices are not equally available to decision-makers—decisions are always conditioned by what we "know" and "value" (Ostrom, 1999; Berkes et al., 2003).

The bottom line is that we will not get closer to an integrated framework unless we treat practices as something more than a stylized component in our models—in which the human dimension is a black box where the behavior and the cultures and contexts that sustain farm practice are assumed rather than measured and modelled. Only when we can relate changes in practice and outcome more directly will farmers and policymakers begin to see how alternative practices can achieve similar levels of productivity with less environmental impact. This argues for more intensive place-based research to flesh out the time-varying nature of social and ecological change. There will still be a need for the geographic breadth of CEAP-like surveys. But to understand adaptive change, we must understand the reciprocation between human and natural systems over time. Groffman and colleagues provide us with a good starting framework. If we could add one important dimension to what they propose, it is that we need to make room for the analysis of adaptive behavior in this enterprise. Culture and time dimensions are essential.

References

Berkes, F., J. Colding, and C. Folke. 2003. Navigating social-ecological systems: Building resilience for complexity and change. Cambridge University Press, Cambridge, United Kingdom.

Carpenter, S.R., and P.H. Carpenter. 1999. Ecological and social dynamics in simple models of ecosystem management. *Conservation Ecology* 3: 4.

Diamond, J.M. 1997. Guns, germs, and steel: The fates of human societies. W.W. Norton, New York, New York.

Gunderson, L.H., and C.S. Holling. 2002. Panarchy: Understanding transformations in human and natural systems. Island Press, Washington, D.C.

Kinzig, A.P., S.W. Pacala, and D. Tilman. 2001. The functional consequences of biodiversity: Emipirical progress and theoretical extensions. *Monographs in Population Biology* 33: i-xxvi, 1-365.

Levin, S.A. 1999. Fragile dominion: Complexity and the commons. Perseus Books, Reading, Massachusetts.

Makuch, J., S.R. Gagnon, and T.J. Sherman. 2004. Agricultural conservation practices and related issues reviews of the state of the art and research needs: A conservation effects assessment bibliography. National Agricultural Library, Beltsville, Maryland.

McDonnell, M.J., and S.T. Pickett. 1993. Humans as components of ecosystems: The ecology of subtle human effects and populated areas. Springer-Verlag, New York, New York.

National Science Foundation. 2002. Long-term ecological research twenty-year review. http://intranet.lternet.edu/archives/documents/reports/20_yr_review.

Ostrom, E. 1999. Self-governance and forest resources. Occasional Paper 20. Center for International Forestry Research, Jakarta, Indonesia.

Redman, C.L. 1999. Human impact on ancient environments. University of Arizona Press, Tucson.

Vitousek, P.M., J.D. Aber, R.W. Howarth, G.E. Likens, P.A. Matson, D.W. Schindler, W.H. Schlesinger, and D.G. Tilman. 1997. Human alteration of the global nitrogen cycle: Sources and consequences. Ecological applications: a publication of the Ecological Society of America 7:737-750.

What science can and cannot provide policymakers

Otto Doering

Information and its context

In their chapter, "Ecosystem Services in Agricultural Landscapes," Peter Groffman and colleagues summarize the state of our knowledge about how to account for environmental effects and the steps that must be taken to allow us to do that better. What will be critical is the ability to measure ecosystem services and convey at least their relative value. In addition, such goods and services must be those that the public values and identifies with if they are to enter the decision process for setting objectives and allocating resources. Hopefully, the ability to do this can be the basis for answering the critically important "so what" questions that are posed about a public program or activity.

Measurement is critical to the "so what" question in that it has to be accurate enough for the task at hand—in some cases only relative magnitudes may be required. It may or may not matter whether the answer is through monitoring or modeling. What will be critical is that the public and policymakers understand where the science is at in its ability to inform through measurement and assessment and also to understand the degree of validity that can be ascribed to the information supplied.

Improved information about ecosystem services and how those services arise from an agricultural landscape can provide a most useful tool for the design, implementation, and management of conservation programs. Included in "policymakers" are those who, through their influence on the design and implementation of programs responding to policy directives, have in fact a strong influence on policy. Thus, program managers and even implementers influence the course of policy ex post as well as ex anti. Policy impact goes well beyond the person identified as the policymaker.

When one gets closer to the actual management and implementation of a program, the information necessary may be different. It may have to be more specific; or it may have to have enhanced validity to be useful in such requirements as adaptive management. It is the adaptive management role where timely provision of information and feedback from programs appears to be most deficient, largely because adaptive management information requirements are not usually conceived of as an essential part of the program to begin with (Soil and Water Conservation Society, 2006). A simple but formal template for reviewing progress and stimulating consideration of program change to better achieve existing or modified objectives is a must for future programs. It is here that the information requirements from science will have to be tailored specifically to the activity at hand and objectives of the program.

In considering what science can (or should) provide policymakers, context becomes important. In fact, context can be as important as or more important than relevant scientific information or specific ecosystem outcomes. The public's perception and the resulting political context are usually the drivers of what the public is interested in, often irrespective of the ecosystem services that might be provided. Costanza's valuation of world ecosystem services was an intriguing exercise, but it has not resulted in the public or government seriously evaluating ecosystem services or significantly increasing the value placed on those services in the decision framework for activities that affect ecosystems positively or negatively (Costanza, 1997). This may occur in the future, but changing the public perception of what is important is a long-term process.

Simple answers, complex questions

Another difficulty that science faces in providing information to policymakers is that the public and policymakers want the equivalent of the canary in the coal mine. That is, they desire a simple, discrete signal that things are okay or not okay. If the canary is alive, we are okay. If the canary dies, we had better get out of here. From the ecosystem standpoint, policymakers would like to have discrete choices with discrete outcomes. Improvement in our understanding of the science may make such judgments harder rather than easier as we realize that things are more complex than we first supposed. Such questions as how much less nitrogen flowing down the Mississippi River will eliminate gulf hypoxia or what reduction in carbon emissions will halt that factor's contribution to climate change represent where policymakers would like science to be. The truth is that in so many cases science just cannot be there.

It is especially difficult in this context to provide good information that also recognizes spatial and temporal tradeoffs. With nitrogen leaking from agricultural systems into water systems, we know that soil systems may continue to give off excess nitrogen long after excess nitrogen fertilizer has been applied. This seeming lack of spatial and temporal cause and effect makes life extremely difficult for policymakers and confuses the public.

One way to provide useful input to policymakers, even considering these complex questions, is through pilot projects. Piloting government programs used to be standard operating procedure for many agencies. This is now seen as a costly and not very productive step. I would argue that, while the Conservation Security Program was supposed to be rolled out from the start as a national program. the reduced, budget-driven, and pilot nature of the rollout so far has improved the design, administration, and potential impact of the program for subsequent years in those areas not yet included in the program. For much of what we do on the land, especially given our need to understand ecosystem response better, pilot projects give us at least a few reality checks. They can be drawn on in an illustrative way that is meaningful to policymakers and the public.

Selling public goods

Policymakers are not only conceptualizing, designing, and implementing programs dealing with the landscape, but they also are having to calculate what the public is willing to pay for. We have seen over the last 50 years a shift in what the public considers to be public goods—goods that benefit all—are nonexclusive to individuals and could not be paid for by a single individual. Years ago, education and conservation were considered to be public goods. Today, the mantra is that only students benefit from education, so they should pay for it.

Conservation programs come increasingly under more stringent benefit-cost analyses to demonstrate present value. The implicit question here is this: Should the public be asked to pay for something today when the returns may be far in the future—maybe even give proportionally more benefit to a future generation that did not have to pay the cost? There was more willingness for previous generations to undertake this kind of expenditure for education and conservation than there appears to be today. This puts today's policymaker in a difficult situation when proposing a long-term program with current expenditures and a forward-reaching stream of benefits, no matter how good the science.

Summary and Conclusions

Policymakers are confronted with a difficult public decision environment that is not fully dealt with through a better knowledge of ecosystem services and the behavior of ecosystems across the landscape. Some barriers to good policy cannot be solved with sound science. Science can be most helpful in aiding both the initial decision to commit, through understanding the necessity for and the impact of policies, and in the adaptive management portion of a program once it gets underway. Science is critical in providing information to answer the "so what" question for a policy, but scientists also need to learn to identify the "so what" questions important to policymakers, the public, and decisions to go ahead with a program. Politically, policymakers must be interested in the big picture for their constituencies, and they always must find out what the impact will be in their respective district. Science has to be able to respond appropriately with knowledge and information.

References

Costanza, R., R.d'Arge, R. De Groot, S. Farver, M. Grasso, B. Hannon, K. Limburg, S. Naeem, R.V. 'ONeil, J. Paruelo, R.G. Raskin, P. Sutton, and M. van den Belt. 1997. The values of the world's ecosystem services and natural capital. *Nature* 387:253-260.

Soil and Water Conservation Society. 2006. Final report from the Blue Ribbon Panel conducting an external review of the U.S. Department of Agriculture Conservation Effects Assessment Project. Soil and Water Conservation Society, Ankeny, Iowa.

Part 2

Methods for environmental management research at landscape and watershed scales

New approaches to environmental management research at landscape and watershed scales

G. Philip Robertson
L. Wesley Burger Jr.
Catherine L. Kling
Richard Lowrance
David J. Mulla

Agriculture is the world's largest industry, and as a landscape-based enterprise, it has an environmental impact that extends well beyond the borders of individual fields. Water, nutrients, dust, pests, pollen—whether managed intentionally or unintentionally—leave fields and farms for points downstream and downwind, affecting organisms, water quality, and air quality at locations sometimes far distant from their points of origin. Such effects have been a defining feature of agriculture since its early days. Indeed, early atmospheric methane increases have been linked to the onset of lowland rice cultivation more than 2,000 years ago (Intergovernmental Panel on Climate Change, 2002), and carbon dioxide increases since 1750 have been caused in part by the onset of widespread land conversion to cropland following European expansion (Wilson, 1978). More

recent effects, ranging from nutrient exports to groundwater to the regional delivery of windborne, micron-sized particulates, are well documented (National Research Council, 2000; Aneja et al., 2006).

Arguably, agriculture depends upon landscapes as much as agriculture affects them. Natural predators of crop pests, for example—important even where pesticides are used and crucial elsewhere—are able to colonize crops to feed on pests only where landscape features such as woodlots and unmanaged fields provide food, cover, and other resources that allow them to persist during those parts of the year that crops are unavailable (Landis et al., 2005). Soybean aphids in the Great Lakes region are readily controlled by Coccinellid beetles in landscapes with woodlots that provide overwintering habitat and successional fields that provide early flowering native plants for spring food. Likewise, crops depend upon landscape heterogeneity for other services: for pollinators, for climate moderation, for irrigation recharge, for windbreaks, and for mitigating the downstream effects of agriculture that would otherwise be much more severe.

These latter effects are perhaps the best studied and most purposefully managed landscape attributes. First-order streams and wetlands can transform nitrate leached from agricultural fields to more inert forms of nitrogen, such as nitrogen gas (Lowrance, 1998). Riparian buffer strips trap overland flow and its transport of phosphorus and sediment to surface waters (Richardson and Qian, 1999). Successional fields, including Conservation Reserve Program (CRP) land, provide habitat for birds and other wildlife that would otherwise be more completely displaced by agricultural conversion (Ryan, 2000; Farrand and Ryan, 2005; Johnson, 2005).

In short, agriculture is as much as ever, and perhaps more so today, a landscape enterprise. And as we move into an era in which ecosystem services from agriculture are tabulated and valued (Robertson and Swinton, 2005; Swinton et al., 2006), landscape involvement and management will become ever more important. A majority of the noncommodity services provided by agriculture involve landscape elements and landscape-level processes. Clean air, clean water, biodiversity, wildlife, and visual amenities are but a few examples.

Why, then, are agricultural landscapes under-studied with respect to these processes? A variety of reasons have been cited; they range from paradigm limitations to social and scientific barriers (e.g., NRC, 2003; Robertson et al., 2004). Underlying most reasons, however, are the methodological: Too few studies have employed appropriate methodologies at the appropriate scales to provide the comprehensive knowledge needed to manage landscapes effectively for the full suite of services they can provide.

While we believe this insufficiency is regrettable, we also believe it can be overcome, and we elaborate on this belief in the pages that follow. Our intent is to provide, first, an overview of traditional field-scale approaches to environmental management and research in agriculture, highlighting both successful examples and their general limitations; second, to describe what we view as the three main challenges for incorporating landscape methodologies into the present research portfolio; and third, to define and describe what is needed in the way of new tools, approaches, and research to overcome the barriers that presently inhibit a landscape orientation and that could help address important and otherwise recalcitrant environmental problems associated with agriculture.

Traditional approaches to management and research

Successful research strategies

Traditional approaches to studying and managing environmental issues in agriculture most often are field-based, sometimes farm-based, and least often watershed-based. Field-scale approaches have proved especially useful for defining the range of management options available for solving specific field-level problems. Nutrient management and wildlife conservation are research areas that illustrate the range of successful approaches for conducting field-scale research.

The U.S. approach to nutrient management research and extension for more than 50 years has been based on statistical analyses of small-plot trials in which crop responses to several fertilizer or manure treatments are tested in a randomized and replicated experimental design. Such experiments are traditionally conducted at agricultural

experiment stations, usually located on level, relatively homogeneous soils representing major soil types. Farmers and agricultural industry representatives are invited to view the results of these trials at field days and experiment station open houses. Results from those experiments are used to develop fertilizer and manure recommendations and are often published in leading research journals and experiment station bulletins. These experiments have shown the value of soil testing for fertilizer recommendations and manure testing for manure management. They also have provided valuable information about impacts of rate, timing, method of application, and formulation of various fertilizer products on crop yield and nutrient loss.

In recent years, plot-based nutrient management research has been extended to include on-farm sites. On-farm variety and tillage trials have long been used to bridge plot-level, station-based research and farmer adoption. On-farm nutrient management research likewise allows in-depth investigation across a wider variety of soils and landscape positions, questions normally limited in station-based research. While experiments at the plot scale are statistically rigorous, straightforward to design and implement, and relatively inexpensive, researchers, industry professionals, and farmers have increasingly become concerned about the difficulty of extrapolating results from small-plot experiments across the broader landscape. On-farm research provides the potential for broader extrapolation, especially with new precision-based technology. Together, plot-based experiments and on-farm research have been hugely successful in providing solutions for specific field-based nutrient management problems.

In contrast to nutrient management research, experiments using tightly controlled, small-plot designs are rarely useful in wildlife science. Although experiments at the plot scale may be appropriate for studies of vegetation or even insect response, wildlife populations typically respond to land use at much larger scales, requiring alternative experimental approaches. Moreover, what constitutes an appropriate spatial scale varies among species with different life-history characteristics. Consequently, much of the evidence for wildlife benefits of conservation practices comes from studies conducted at the practice and field scales in which management of the non-cropped portion of the landscape is varied, with replication across relatively homogenously managed practice units or fields. With these designs, researchers often attempt to control for extraneous sources of variation by blocking or pairing on farm, landowner, location, landscape context, or management regime.

Simultaneous use of controls and treatments in designed experiments allows strong inference through the establishment of both necessary and sufficient causation. In contrast, the identification of appropriate no-treatment "controls" is often problematic in field- or farm-level conservation studies. The reason: It is not always clear whether a control should represent status quo conditions, agricultural production in the absence of the conservation practice, alternative implementations of the conservation practice, or baseline wildlife populations in the natural community for which the conservation practice is a surrogate (e.g., remnant native prairies versus CRP fields).

Random assignment of treatments and controls to experimental units is a hallmark of the scientific method and guards against systematic error, spatial autocorrelation, and subtle researcher bias. But unlike small-plot experiments that are often conducted on agricultural experiment stations, farm- and landscape-scale studies of conservation practices are most often carried out on private working land. As such, the researcher rarely has control over treatment assignment. For instance, most field-, farm-, and landscape-scale studies of wildlife responses to conservation practices are conducted as observational studies where previously treated experimental units (e.g., CRP fields) are randomly selected from the population of units to which the researcher wishes to make inference (e.g., all the CRP fields in a watershed, county, state, or region). Regrettably, convenience sampling too often is substituted for probabilistic sampling, which substantially impairs inferential strength.

Four lines of evidence demonstrate success of a particular wildlife conservation practice (after Ryan, 2000): (1) Occupancy of the practice by the focal wildlife species; (2) high population abundance in the conservation cover relative to alternative habitats, in particular those land uses or land covers that the conservation practice replaces (e.g., cropland); (3) reproductive success sufficient for positive population growth (i.e., λ greater than 1); and (4) positive population growth (or reduced decline) after initiation of the practice.

Despite all of its experimental limitations, extant field-, farm-, and landscape-scale wildlife research documents show overwhelming benefits of conservation practices implemented under U.S. Department of Agriculture (USDA) conservation programs, particularly the CRP (Burger, 2005; Clark and Reeder, 2005; Farrand and Ryan, 2005; Johnson, 2005; Reynolds, 2005). Moreover, those studies demonstrate that accrued benefits vary among species in relation to physiographic region; season; conservation practice and cover; time since establishment; disturbance regimes; patch size, shape, and distribution; and landscape context.

The need for new approaches

Traditional field- and farm-scale management approaches cannot in themselves provide relief from agroenvironmental problems that are expressed at regional to continental scales. As successful as many of the traditional approaches have been, they—and especially the ones that focus on increasing overall production rather than on more efficient production—suffer from limitations that restrict their effectiveness or implementation at larger scales. Mainly, this is due to factors that cannot be addressed with research designed to answer single, narrowly defined questions in the context of individual, field-based management practices.

To address problems that manifest themselves at larger scales, a different approach is needed. Experience suggests that effective solutions will share some combination of the following attributes:

- A systems orientation that balances multiple aims against known tradeoffs. It is well and good to find a specific solution to a specific problem, but ensuring that the solution does not create problems elsewhere in the landscape requires a systems orientation, especially when effects are indirect or offsite. A systems orientation also allows multiple aims to be balanced and synergies optimized.
- Geographic scalability, whereby positive effects at one scale do not disappear at larger scales. Solutions developed at small scales may not be appropriate or even applicable at larger scales. And spatial heterogeneity and

landscape complexity can affect the efficacy of many management practices and in some cases—such as wetland nitrate and phosphorus attenuation—provide solutions not available at smaller scales.

- Socioeconomic considerations that provide for solutions that can be implemented using realistic incentives at an acceptable economic and social cost. Solutions that fail to consider social acceptance or economic viability at multiple scales risk irrelevance due to lack of acceptability and thus implementation. Solutions obvious to the biophysical scientist or agronomist may have hidden social costs and unforeseen barriers to adoption.

- Long-term responses are included, such that climatic, social, and other factors that change on a years-to-decades time scale can be evaluated and against which management impacts can be clearly distinguished from impacts due to long-term trends in the biophysical environment, such as regional climate change. Many environmental attributes change slowly or are affected by major events at infrequent intervals. Without long observation periods that allow changes to be assessed and climatic, population-outbreak, and other infrequent events to be captured, it is difficult to develop informed, robust solutions.

Challenges to adoption of new approaches

Three main challenges confront adoption of landscape-level approaches to environmental research in agriculture: (1) The systems approach is difficult and expensive; (2) regionalization requires extensive sampling and modeling; and (3) the inclusion of socioeconomics requires a new research paradigm.

The systems challenge

Examples of systems approaches to environmental problems in agriculture are few, largely because of the geographic scale and information needs of a systems approach. Those needs make the approach time-consuming and expensive, but ultimately more effective. Two examples illustrate the challenge: Nutrient management and wetland restoration.

Nutrient management. Notable exceptions to the pattern of small-scale approaches to reduce the transport of nutrients, sediments, and greenhouse gases from agriculture are the Management Systems Evaluation Areas (MSEA) and the Conservation Effects Assessment Project (CEAP) programs. The goal of the MSEA program, funded by USDA in the 1990s, was to develop and promote agricultural management systems that reduce the impact of farming on groundwater and surface water quality. MSEA sites (plot, field, and small watershed scales) were located in Ohio, Missouri, Minnesota, Iowa, and Nebraska (Ward et al., 1994). Extensive evaluation of the water quality impacts of farming systems were conducted at those sites, where numerous best management practices (BMPs) were evaluated for their relative effect on water quality. Water quality modeling was used to predict that water quality would improve at watershed and regional scales with reduced applications of phosphorus or nitrogen fertilizers and increased adoption of soil conservation practices.

In contrast to MSEA, the CEAP program is designed explicitly to study the relationships between agricultural management practices and water quality at the watershed scale and, in particular, to evaluate the effectiveness of BMP implementation in select watersheds with a long record of water quality monitoring data (Mausbach and Dedrick, 2004). These studies are designed to address the effectiveness of BMPs for soil erosion control and nutrient management over a wide range of soil, landscape, climatic, and land use characteristics. CEAP studies will also be used to test the accuracy of computer model predictions on the effectiveness of BMPs. Finally, the studies will be used to evaluate the impacts of BMPs on wildlife populations and on soil and air quality.

By considering the impact of alternative BMPs on a variety of system attributes and response variables, CEAP is more systems-oriented than most such programs. Nevertheless, it is incomplete. What constitutes a complete systems approach for nutrient management issues? A complete approach should study the system at the spatial and temporal scales for which it is possible to clearly identify relevant inputs, processes, outputs, feedback loops, nonlinearities, complexities, recovery patterns or resilience, and external controls. It should recognize that agricultural management is affected by farm, energy, financial, environmental, and health policies. It should involve the study of production characteristics; landscape features; farm manager attributes; variability in climate and hydrology; environmental impacts on soil, water, air, and health; and such issues as biodiversity, connectivity, and wildlife habitat.

Systems approaches recognize complexity. For example, a given BMP does not have the same effectiveness for improving water quality across all soil types, landscape positions, climatic regions, or management systems. A sediment BMP differs in effectiveness depending upon slope steepness, distance from a surface water body, and frequency of intense storms. A nitrogen BMP varies in effectiveness in response to such factors as soil organic matter content, amount and timing of fertilizer applied before the BMP was implemented, manure management practices, and extent of subsurface tile drainage. These types of interactions involve complexity.

Systems approaches recognize nonlinearity. For example, the effectiveness of BMPs on water quality may depend upon thresholds or critical values. Reducing phosphorus fertilizer application rates, for instance, may have little impact on water quality if soil phosphorus levels are excessive, but the same reductions may have a dramatic impact if implemented on another soil with moderate soil phosphorus levels.

To complicate matters further, the effectiveness of a nitrogen BMP may depend upon both complexity and feedback loops. For example, the effect on water quality of reducing nitrogen fertilizer application rate may depend upon the amount of crop residue left behind for soil erosion control and the type of tillage practiced. Greater amounts of residue may immobilize more nitrogen, thereby reducing leaching losses. The reduced tillage practices associated with increased crop residue coverage may, however, lead to greater infiltration. Greater infiltration may increase the risk of nitrate leaching. So, reduced tillage systems may either increase or decrease the effectiveness of nitrogen BMPs, depending upon the overall impact of tillage on immobilization versus infiltration.

As another example of complexity and feedback loops, consider controlled drainage, which is promoted as a method for reducing nitrate losses to surface waters. This reduction typically comes at the expense of increased emissions of nitrous oxide, a greenhouse gas. From a systems perspective, the increased greenhouse gas emissions could lead to further global warming and increased precipitation, which could potentially offset the benefits of direct reductions in nitrate emissions to surface waters from controlled drainage (Robertson, 2004).

Finally, systems approaches recognize the larger policy context in which systems outputs are interpreted. During the 1980s, farm policy was heavily influenced by the goal of increased crop production and sustainability of soil resources. Soil erosion control was discussed in the context of keeping soil losses below soil tolerance values so that there would be no long-term reductions in crop productivity. Currently, soil erosion control is being discussed in terms of policies whose goal is to maintain crop productivity while protecting environmental quality at a level assessed by aquatic goals [e.g., total maximum daily loads (TMDLs), concentrations of total suspended solids (TSS), substrate embeddedness, and vigor of submerged aquatic vegetation]. In this context, it makes little sense to find management solutions that reduce erosion to soil tolerance values because more dramatic reductions are generally needed. The larger reductions needed are producing a wider range of management options for soil erosion control than conservation tillage, including riparian buffer strips, cover crops, alternative cropping systems, and land retirement.

Wetland restoration. One of the most important changes in agricultural landscapes that has taken place as a result of conservation programs in the last decade is the restoration of wetlands. This has been accomplished through various USDA and other programs. Most wetlands restored through these programs are former wetlands in which hydrology had been altered to make crop production feasible. Although not generally applied to restored agricultural wetlands, functional assessment techniques are available to determine when wetland functions have been restored (Brinson et al., 1995; Brinson and Rheinhardt, 1996). Functional assessment of wetlands has been driven by section 404(d) of the Clean Water Act, which has limited the draining and filling of wetlands. Functional assessment techniques have been developed to guide wetland mitigation. Those techniques are also applicable to voluntary, incentive-based wetland restoration.

Functional assessment techniques depend upon comparison of wetland attributes to those of a reference wetland (Brinson and Rheinhardt, 1996; Findlay et al., 2002). Lists of wetland functions (e.g., Table 1) have been developed for a variety of wetland types. Those assessments have been driven by both wetland protection and mitigation. One major impediment to understanding the functional restoration of wetlands is that functional assessment depends in large part upon the comparison of restored wetlands to reference wetlands. In many agricultural landscapes, there are no reference wetlands available because of changes in hydrology and land surfaces (geomorphology) associated with the original conversion of wetlands to arable farmland. Without reference wetlands, functional assessments must depend upon a few key factors rather than a suite of factors.

One of these key factors is wetland hydrology, which illustrates the danger of using one factor alone to assess wetland function. Examples abound of hydrologically restored but disturbed wetlands that have been overtaken by exotic invasive plants, such as Brazilian pepper, purple loosestrife, and reed canary grass; as a result, those wetlands have few functional attributes of the original wetlands they were intended to restore (Ferriter, 1997; Zedler and Kercher, 2004).

Biogeochemical budgets can also be used to assess functionality. Such budgets provide an integrated measure of system performance and thus can be used to evaluate the system-level response to restoration efforts. But long-term monitoring and large amounts of resources are necessary to develop hydrologic budgets, and the expense is even greater when nutrient and sediment budgets are pursued. Nevertheless, wetland restorations and creations have been evaluated using input-output budget techniques on a scale ranging from a few hectares or acres to many square kilometers or miles (e.g., Kovacic et al., 2000; Vellidis et al., 2003).

The purposes of wetland restoration efforts funded under USDA's Conservation Reserve Enhancement Program (CREP) may differ, even in states with similar problems. In the U.S. Corn Belt, for example, the Iowa CREP specifically focuses on reduction of nitrate from tile drains. Wetland siting is based on the primary criterion of controlling nitrate movement from drainage systems (Figure 1). The effectiveness of those wetlands are evaluated by following long-term changes in watershed nitrate transport. The greatest potential benefits for nitrate mass reduction will be in extensively row-cropped and tile-drained areas of the Corn Belt where the nitrate loads are highest (Crumpton, 2000, 2005). But removal rates are not only driven by nitrate loading; hydrologic loading and water residence times are important as well (Kovacic et al., 2000).

In contrast, the Illinois CREP is focused on a wider array of goals, including sediment load reduction; nitrogen and phosphorus load reductions; increased populations of waterfowl, shorebirds, and threatened and endangered species; and increased native fish and mussel stocks. Establishment of wetlands for a wider range of purposes that are more consistent with general stream restoration goals is likely to increase the potential of restoration sites. But evaluation of programs with more complex criteria is more difficult than the evaluation of those with more simple goals.

The regionalization challenge

The second challenge facing incorporation of new approaches is that of regionalization, which requires extensive sampling and extrapolation that, for a variety of reasons, can be difficult. The management of nitrate, sediment, and wildlife illustrate these issues.

Management of nitrate on a landscape scale. High concentrations of nitrate in tile-drain systems lead to flow-weighted concentrations of nitrate above 10 milligrams of nitrate-nitrogen per liter in many streams draining the agriculturally important region of the Corn Belt. Artificial subsurface drainage causes water to move through the root zone much faster than in undrained land. Because of this increased rate of movement, nitrate is leached to the drain system and then moves out to receiving waters. This phenomenon occurs in most environments where agriculture is conducted on drained soils: The Southeast (Lowrance et al., 1984), the Lake Erie Basin (Calhoun et al., 2002), Ohio (Tan et al., 2002), and Quebec (Elmi et al., 2004). Although most nitrate in agricultural landscapes is derived from fertilizers, it

Table 1. Functional characteristics of riverine wetland classes used to assess wetland restoration success.

Class	Functional attribute
Hydrologic	Dynamic surface water storage
	Long-term surface water storage
	Energy dissipation
	Subsurface storage of water
	Moderation of groundwater flow or discharge
Biogeochemical	Nutrient cycling
	Removal of imported elements and compounds
	Retention of particulates
	Organic carbon export
Plant habitat	Characteristic plant communities
	Characteristic detrital biomass
Animal habitat	Spatial structure of habitat
	Interspersion and connectivity
	Distribution and abundance of invertebrates
	Distribution and abundance of vertebrates

Source: Brinson et al., 1995.

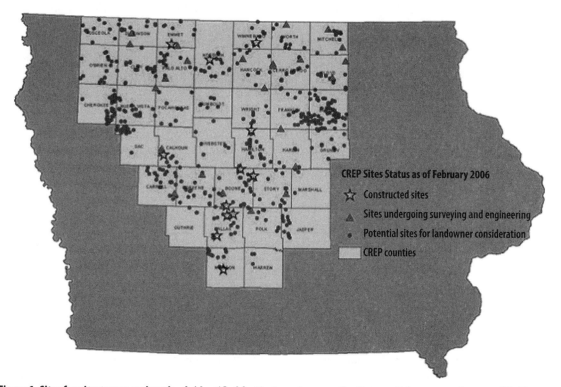

Figure 1. Sites for nitrate removal wetlands identified for the Iowa Conservation Reserve Enhancement Program (CREP). Source: Iowa Department of Agriculture and Land Stewardship, Water Resource Bureau (http://www.agriculture.state. ia.us/CREP.htm).

can be derived from any nitrogen source, includ-ing soil organic matter, manure, and atmospheric deposition.

The effects of excess nitrate on environmental health are generally manifested through enhanced algal growth in downstream water bodies, even-tually leading to lowered levels of dissolved oxy-gen and other signs of eutrophication, depending upon nitrogen:phosphorus ratios in freshwater or nitrogen:phosphorus:silica ratios in coastal waters.

Management of nitrate is tightly tied to man-agement of the nitrogen cycle, which has myriad biological and chemical reactions that affect nitro-gen oxidation state and transportability. Because different landscape components can be structured and managed to encourage these reactions, there are many landscape-scale management options for nitrogen in general and nitrate in particular. Measurement of the effects of these landscape management options can range from sophisti-cated techniques tied to gaseous evolution, to techniques to understand the movement and retention time of water, to techniques to under-stand the relative areas of cropland versus wet-land. In tile-drained landscapes, measurement of changes in nitrate transport in response to land-scape and agricultural management is relatively straightforward.

Although measuring nitrate levels in water is straightforward, interpreting the effects of land-scape management or adoption of nitrogen-con-serving practices on a landscape scale is difficult. Numerous studies, for instance, have shown that in the Corn Belt a switch from fall applica-tion of anhydrous ammonia to spring application of other nitrogen sources (e.g., urea ammonium nitrate) will lead to 15 to 20 percent reductions in nitrate concentrations (and loads) in tile drainage water (Randall and Sawyer, 2005). But document-ing the change in nitrate concentration and load on a landscape or watershed scale for this rela-tively simple nitrogen conservation technique has been difficult, largely because of the inherent diffi-culty of paired watershed studies and comparing watersheds over time.

Jaynes et al. (2004), for example, found that in a 400-hectare (1,000 acres) treated watershed, where most farmers had switched to spring nitro-gen application (late spring nitrate test), average annual flow-weighted concentrations were lower

than in two control watersheds, where farmers did not switch from fall anhydrous applications. The study was carried out for five pre-treatment and four post-treatment years (1992-2000), and differences between the treated watershed and the two controls were only evident in the last two years of treatment (1999 and 2000) (Table 2). Study results were complicated by the finding that one of the control watersheds consistently had the lowest concentrations in the pre-treatment period and by the finding that nitrate concentrations in the treated watershed only declined about 1 mil-ligram of nitrate-nitrogen per liter during the four post-treatment years compared to the pre-treat-ment periods. Nevertheless, Jaynes et al. (2004) concluded that the late-spring-nitrogen-test treat-ment resulted in a 30 percent or greater reduc-tion in nitrate-nitrogen in the drainage water in the last two years of the treatment period (Figure 2), with no reduction in yields. Although results showed that the late-spring-nitrogen-test-treated subbasin had lower nitrate concentrations, the measurements showed the difficulty of demon-strating the efficacy of nitrogen-conserving prac-tices on a watershed scale. In two years of the treatment period (1997 and 1999) actual nitrogen reductions on the treated watershed were small or nonexistent. In two years (1998 and 2000), the fertilizer rates were greatly reduced [by 46 and 73 kilograms (41 and 65 pounds) of nitrogen per hectare per year]. Given the difficulty and expense of these types of watershed-scale studies, it is not surprising that there are few other published studies of this sort.

Phosphorus and sediment loads at landscape scales.
Efforts to assess BMPs to reduce phosphorus and sediment loads from watersheds illustrate differ-ent approaches to watershed assessment. Tradi-tional approaches typically involve water quality monitoring at the mouth of the watershed, before and after implementation of BMPs within the watershed. This approach is limited in its ability to study cause-and-effect relationships because water quality monitoring data reflect cumulative impacts of large areas with different landscape features, management practices, vulnerabilities to pollution export, and climatic patterns. Further-more, long-term water quality data sets (greater than 10 years) are typically needed to identify trends in water quality. An example of the latter is

Table 2. Flow-weighted average annual nitrate-nitrogen (NO3-N) concentrations in water draining Iowa watershed sub-basins in which fields were nitrogen-fertilized as usual (control 1 and 2) or nitrogen-fertilized on the basis of late spring nitrate tests (LSNT) beginning in 1997 (TR1).

Subbasin	Control 1	Control 2	LSNT
1992 (mg NO$_3^-$-N L^{-1})	9.9	13.7	12.5
1993 (mg NO$_3^-$-N L^{-1})	8.2	9.7	9.2
1994 (mg NO$_3^-$-N L^{-1})	9.2	10.2	8.9
1995 (mg NO$_3^-$-N L^{-1})	13.1	16.7	16.0
1996 (mg NO$_3^-$-N L^{-1})	14.0	15.4	15.6
1997 (mg NO$_3^-$-N L^{-1})	8.4	13.1	10.8
1998 (mg NO$_3^-$-N L^{-1})	11.1	14.0	10.2
1999 (mg NO$_3^-$-N L^{-1})	15.8	16.5	11.7
2000 (mg NO$_3^-$-N L^{-1})	16.5	15.1	11.0

Source: Jaynes et al., 2004.

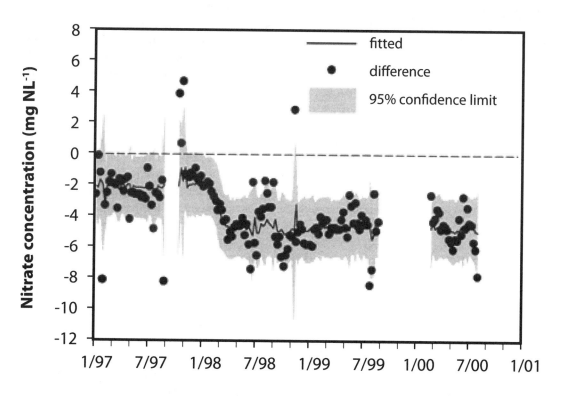

Figure 2. Change in flow-weighted drainage nitrate levels from a watershed subbasin in which fields were nitrogen-fertilized based on late season nitrate test (LSNT) relative to levels from a business-as-usual subbasin. Source: Jaynes et al., 2004.

a study conducted by Richards and Baker (2002) on four watersheds in Ohio. They studied water quality data from 1975 to 1995 using analysis of covariance, with time and seasonality as covariates. Significant reductions were observed in total phosphorus and total suspended solids, but not nitrate-nitrogen.

A second approach to watershed regionalization is water quality monitoring upstream and downstream of the area where BMPs are implemented. Water quality downstream of BMPs can be compared with water quality upstream to determine if there have been any improvements. This approach is of limited value, however, if the upstream monitoring station collects water from a very large area because it will be difficult to detect small changes in water quality due to implementation of BMPs downstream.

A third approach is multiyear monitoring of multiple watersheds where BMPs have been implemented. A challenge, however, is normal variability in river flow. It is difficult to separate the influences of flow variation due to climatic variability from the effects of BMPs. Davie and Lant (1994), for example, studied the impact of CRP implementation on sediment loads in two Illinois watersheds. They found that CRP enrollments on 15 and 27 percent of cropland reduced estimated soil erosion rates by 24 and 37 percent, respectively, but sediment loads at the mouths of the watersheds declined by less than 1 percent. They attributed those small overall impacts to poor targeting of CRP to land in close proximity to streams and to a time delay in sediment transport from the field edge to the mouth of the watershed.

The most rigorous approach to watershed-based regionalization involves paired watershed comparisons, as in the Jaynes et al. (2004) nitrate example above. Udawatta et al. (2002) used field-scale, paired watersheds to study the effects of grass and agroforestry contour buffer strips on runoff, sediment, and nutrient losses on highly erodible claypan soils in northern Missouri. After a seven-year calibration period, grass and agroforestry strips were initiated and found to reduce total phosphorus by 8 and 17 percent, respectively, during the first three years. The contour strip and agroforestry treatments reduced runoff by 10 and 1 percent, respectively, during the treatment period. The contour-strip treatment reduced

soil erosion by 19 percent in 1999, while soil erosion in the agroforestry treatment exceeded the predicted loss.

Birr and Mulla (2005) implemented conservation tillage on 70 percent of the moldboard-plowed acreage for three years in a 1,100-hectare (2,750-acre) watershed in southern Minnesota. No changes in tillage were made in an adjacent watershed. Although these changes resulted in a 40 percent reduction in soil erosion for the treated fields and an estimated 20 percent reduction in sediment load delivered to the mouth of the watershed, statistical comparisons of water quality monitoring data in the treated and control watersheds failed to show any improvements in water quality in the treated watershed. This outcome, according to the researchers, was probably due to (1) the effects of climatic variability, (2) the lag times for transport of pollutants from the field to the watershed scale, and (3) the need for more than three years of water quality monitoring data to identify trends.

Wildlife management at larger scales. The basis of our understanding of wildlife benefits of conservation practices has been derived primarily from studies conducted at the patch, field, and farm scales, and generalizations to larger spatial scales are difficult because of the substantial variability and often contradictory nature of observed responses. Apparent contradictions are frequently attributable to variation in landscape context. Both landscape composition and structure can influence ecological processes such as dispersal, predation, habitat selection, and population performance and, hence, observed responses.

In central North Dakota, for example, daily survival rates of upland-nesting ducks exhibited a curvilinear relationship with patch size, with lowest survival occurring in intermediate-sized patches (Horn et al. 2005). Moreover, landscape composition (percentage of landscape in grassland habitats) altered the functional relationship between daily nest survival and distance to nearest field edge. Those differences in predation patterns may be attributable to effects of landscape composition on predator habitat selection (Phillips et al., 2003), space use (Phillips et al., 2004), and foraging efficiency.

Thompson et al. (2002) hypothesized that effects of habitat fragmentation operate within a spatial hierarchy in which effects at larger scales (regional and landscape) impose constraints on local scale (patch and edge) effects. Consistent with this hypothesis, Stephens et al. (2003) reported that studies where habitat fragmentation was measured at landscape scales were more likely to detect effects on avian nest success than studies measuring fragmentation at local spatial scales. Similarly, Chalfoun et al. (2002) reported that studies conducted at larger spatial scales were more likely to detect functional and numerical responses of nest predators to fragmentation. Horn et al. (2005) concluded that design of conservation cover configurations that meet conservation goals require an understanding of the patterns of nest success and the predation processes that produced observed patterns. Thus, studies designed to evaluate or monitor wildlife responses to conservation management systems need to be explicitly hierarchical in design, executed at larger spatial scales than previously considered, and conducted across a range of landscape compositions and configurations.

A recent CEAP review (Soil and Water Conservation Society, 2006) acknowledged that estimates of environmental effects of conservation programs should be compared to established environmental goals and linked to the ecological context in which the estimated effects occur. Moreover, the panel conducting the review strongly emphasized that simulations and extrapolations must not substitute for on-the-ground monitoring and inventory systems designed to determine if anticipated conservation and environmental benefits are being achieved. The panel specifically endorsed a hierarchical assessment model based on project-level monitoring within the context of regional assessments, and further suggested that a national assessment produced by aggregating valid regional assessments (context-sensitive measures) is more credible and meaningful than generalizing from national to regional scales.

The socioeconomic challenge

Agricultural activities are integrally linked with the nation's ecosystem services and the environmental health of its natural resource base. In previous sections we argued that the complex interactions between land use, water, soil, and air mandate that scientists adopt a systems approach

at the landscape or watershed scale to understand the relationships between agricultural practices and the resulting ecosystem services. To appreciate the scientific need for a paradigm that integrates social and economic influences with the biophysical, it is necessary to emphasize one more feature of the system: The critical role of human actions.

The central United States provides a textbook example of fundamental antrhopogenic change in the hydrologic system and environmental resources in the agricultural landscape. The Mississippi River now carries 15 times more nitrate than any other U.S. river and twice as much phosphorous compared to loads prior to European settlement. More than 95 percent of the original prairie, savanna, and woodland in the Mississippi River watershed has been converted to agricultural use, and farmers actively manage about 75 percent of the land area within the region.

Human behavior has caused these changes. Thus, our attempt to understand the system will be fundamentally incomplete if it fails to incorporate human decision-making factors, such as economic influences, human preferences, and social mores. In short, science that wishes to inform and support policymakers and land use managers must also focus on human behavior and the links from that behavior to land use, ecosystem services, and back again. For example, biophysical scientists may identify the ideal location for reintroduction of a native species, but if economic factors lead to incompatible land uses on neighboring tracts, the reintroduction may be doomed before it begins. Likewise, limnologists may determine that dredging a lake is a cost-effective way to improve lake clarity, but if economic incentives direct farming practices to continue unchanged in the watershed, the lake may soon revert to its murky status.

Perhaps an even more compelling case for a paradigm that explicitly incorporates human behavior comes from the information needs dictated by the current policy environment of agricultural and environmental decisions. Simple observation suggests that the United States has adopted a fundamentally different approach to controlling environmental alteration associated with agricultural activities than in the case of other major industries. Specifically, air and water emissions from industrial and transportation sources are typically controlled by regulations that permit certain levels of emissions, economic fines when emission levels are exceeded, or mandatory caps with trading programs. In contrast, environmental effects associated with farming activities have been addressed primarily through voluntary actions and financial incentives. Major cost-share programs, such as the Environmental Quality Incentives Program (EQIP), the CRP, and the Conservation Security Program (CSP), among others, have pumped billions of dollars into voluntary conservation efforts. With some exceptions [e.g., confined animal feeding operation (CAFO) regulations], there is little meaningful discussion about shifting to a large-scale regulatory paradigm for agricultural conservation policy.

In such a context, federal and state governments, environmental organizations, and all those interested in ecosystem services from agriculture will need information on how best to spend their limited conservation dollars to induce changes on the landscape that most effectively meet their goals. For example, consider a nongovernmental organization or governmental entity interested in purchasing land within a watershed to locate a wetland to improve downstream water quality. A wetland expert may be able to identify ideal hydrologic conditions and spatial features to rank order the sites that would be most effective for this goal. Subsequent action might include purchasing and building a wetland on the first site from the list, then, if funding permits, doing the same for remaining sites. Following this approach, will the nongovernmental organization have done the most that it can for water quality? Not necessarily. Depending upon the cost for each of the sites, including the purchase price and restoration costs, greater water quality improvement might well be had by choosing two sites ranked lower on the list that in combination cost the same as the top site, but together produce greater benefits. The best solution becomes even more complicated if multiple land uses for each site are possible (e.g., differing perennial covers or working land in conjunction with various BMPs) and multiple environmental benefits are sought (water quality and wildlife habitat).

There are numerous other examples of policy questions that only an integrated biophysical-social science paradigm can fully address: What are the tradeoffs between focusing conservation budgets on land retirement (such as the CRP) relative to increasing conservation practices on land still in production? How much more can water quality be improved by targeting conservation dollars on the most beneficial conservation practices and locations within a watershed relative to more equally distributed conservation dollars? How can conservation programs be designed to most effectively improve environmental quality (bidding systems versus uniform payments)? How do different environmental targets (water quality, biodiversity, carbon sequestration) affect the optimal choice of land to enroll in a conservation program? What is the magnitude of the tradeoff between environmental improvement and farm profitability? How effective would a watershed trading program be in meeting water quality goals while simultaneously keeping costs low?

These are just some of the policy-relevant questions that an integrated paradigm can address. Questions of this type can only be considered within an integrated paradigm. Neither a pure social science nor a pure biophysical science approach will do.

Water quality and carbon sequestration. The first link in developing interdisciplinary paradigms that include socioeconomic factors is to develop models that predict adoption of conservation practices and land use change in response to key drivers, such as land characteristics, agricultural productivity, landowner characteristics, and costs and returns to alternative land uses. Indeed, economists and other social scientists have developed a large literature addressing the factors that induce or discourage landowners from adopting various conservation practices and/or land use decisions. For example, Sunding and Zilberman (2000) reviewed much of the economic literature on farmers' adoption of conservation practices and other new agricultural technologies. A complementary literature is quickly developing that considers the key drivers of land use decisions using spatially explicit models in conjunction with geographic information system (GIS) data to consider patterns of urban development, siting of noxious facilities, and spotty rural development. While many of these studies consider the socioeconomic questions relevant to adoption of conservation practices and/or land use change, many do not consider the biophysical consequences of

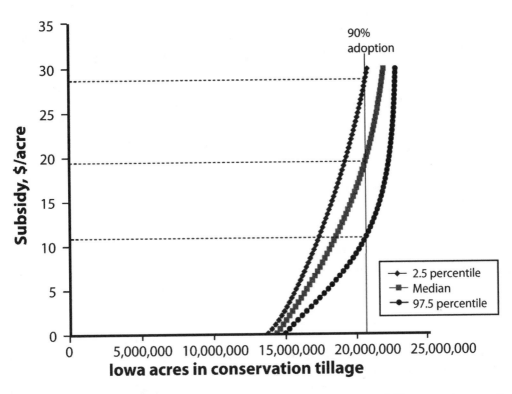

Figure 3. Estimated supply of additional conservational tillage acreage as the per acre subsidy payment increases. Source: Kurkalova et al., 2006.

changes. They fall strictly within the social science paradigm.

Kurkalova et al. (2006), for example, used observed behavior by farmers to quantify the adoption of conservation tillage practices by those farmers. Pairing these data with information on net returns to farming with conservation and conventional tillage, soil and land characteristics, weather patterns, and farmer characteristics, the authors produced a model that predicts farmers' adoption of conservation tillage as a function of its costs and suitability for their location. Using this model, the authors estimated the increased adoption that could be induced by a green payments program that paid a fixed fee for additional adoption of reduced tillage methods. The resulting "supply" schedule of conservation tillage (Figure 3) provides the link between human behavior and land use. Models of this sort can predict how acreage in a conservation practice might change with price and subsidy payment changes, but they cannot predict the consequences of these changes on environmental indicators without an explicit linkage to a biophysical model.

A number of studies incorporate simple biophysical impacts, generally measured in terms of "edge-of-field" environmental gains (e.g., Russell and Shogren, 1993). A few studies have taken the next important step by integrating economic modeling with watershed-based models (Braden et al., 1989; Carpentier et al., 1998; Qiu and Prato, 1999; Ribaudo et al., 2001; Johansson et al., 2004; Petrolia et al., 2005). To demonstrate the development and implementation of a study that incorporates economic behavior in conjunction with a biophysical model, we draw upon two recent studies that focus on conservation policy in the context of water quality and carbon sequestration, two ecosystem services that are likely to drive much of the conservation planning in agriculture in the coming decades. In each of these studies, the authors developed a behavioral model that predicts how landowners or farmers will change their land use (cropping patterns, conservation practices, and land retirement) as a result of changes in economic conditions (such as conservation payments or prices). The behavioral model is directly linked to a biophysical model that predicts how changes in land use will affect the ecosystem service.

Feng et al. (2007) considered the tradeoffs implicit in the design of a conservation program that targets carbon sequestration versus one that targets a water quality indicator, such as soil erosion. They studied only a single land use change: The choice of landowners to enroll in a land retirement program, such as the CRP, but cover a relatively large spatial area (the Upper Mississippi River Basin, which covers portions of seven states). Using the National Resources Inventory (NRI) as their unit of analysis, they linked the Environmental Policy Integrated Climate (EPIC) model to a simple economic model of landowner decision-making about the enrollment decision. There were more than 40,000 cropland "points" in their data set for which they simulated landowner decisions under a variety of hypothetical payment levels.

Findings (Table 3) suggest that policy designed to both reduce soil erosion from the enrollment of land into a CRP-type program and increase carbon sequestration in the soil will face a substantive tradeoff in choosing which of the two environmental indicators to target. For example, assuming a program budget of $500 million, a policy designed to achieve the highest level of carbon sequestration can sequester 2.9 metric tons (3.2 million tons) of carbon by enrolling about 1.5 million hectares (3.75 million acres) of land (Table 3), assuming that policy allows the authority to enroll the land that generates the highest carbon sequestration benefits per dollar spent (similar to the bidding method in the current CRP). This enrollment choice also results in sizable soil erosion control benefits. But if the program were specifically designed to target soil erosion and land was enrolled so that the highest erosion control benefit was achieved, different land would be enrolled at the same budget level. In this case, about 25 percent as much carbon would be sequestered as in the first program, but more than five times as much soil erosion would be eliminated (Table 3). From this sort of analysis it becomes possible to evaluate tradeoffs among different environmental policy decisions and optimize for desired national outcomes.

In a related paper, Feng et al. (2006) used a similar modeling framework to address a different policy question: How does a program that pays farmers to adopt conservation practices on working land (land that stays in agricultural production) affect the cost-effectiveness of a land retirement program? Landowners face an array of choices with respect to conservation programs, including those that provide payments for complete retirement of land (CRP) and those that cost-share or provide payments to adopt conservation practices on working land (CSP). When faced with mutually exclusive alternatives, farmers must make a decision, and their choice whether to enroll in the CRP, the CSP, or not participate in either program will determine the success and cost-effectiveness of each program. Again, an integrated economic and biophysical modeling paradigm allows analysts to consider these questions and provide information that can be used to improve the design of those programs.

An important shortcoming of these examples is that the biophysical model employed, and, therefore, the resulting solutions to the conservation policy choices, measures environmental effects at the edge of each field. It does not capture the interacting effects of land use and offsite water quality. While still a nascent literature, there are also some careful studies that consider the full link from proposed policy enactment to behavior change to in-stream environmental consequences (Khanna et al., 2003; Beaulieu et al., 1998; Kling et al., 2006).

The importance of disproportionalities. There is increasing recognition that nonpoint-source agricultural pollution, especially from sediment and phosphorus, arises from small fractions of the landscape (Gburek et al., 2002). Disproportionalities arise from several factors, including climatic and hydrologic patterns, topographic features, and management factors.

Birr and Mulla (2005) documented such disproportionality in two adjacent Minnesota watersheds in which they surveyed 24 farmers managing 220 fields. Results showed large disproportionalities in applied phosphorus from fertilizer and manure, ranging from 9 to 545 kilograms (8 to 487 pounds) of phosphate per hectare across corn fields, with distributions highly skewed to higher application rates on small fractions of the land base (Figure 4). A phosphorus index approach to evaluate risks of phosphorus loss to surface waters showed that 18 percent of the

Table 3. Distribution of predicted multiple environmental benefits from a uniform subsidy of $500 million for land retirement targeted toward specific benefits.

Target benefit	Benefit distribution			
	Carbon sequestered	Erosion avoided	N runoff avoided	N leaching avoided
Carbon (10^6 metric tons)	3.2	0.8	0.6	1.0
Erosion (10^6 metric tons)	7.4	40.5	14.1	9.7
N runoff (10^3 metric tons)	2.8	5.1	11.7	2.8
N leaching (10^3 metric tons)	10.0	6.4	5.6	30.6

Source: Feng et al., 2007.

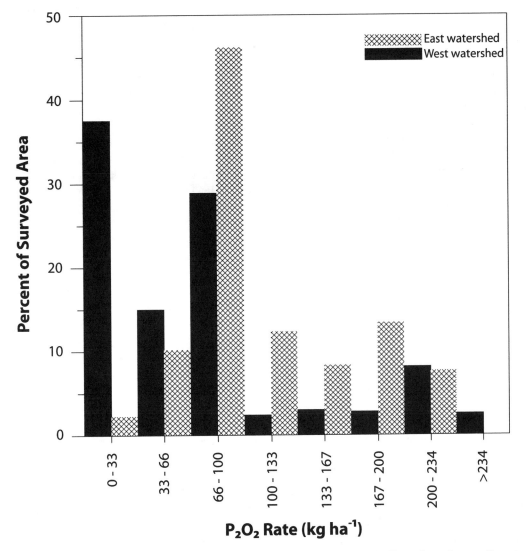

Figure 4. Disproportional distribution of phosphorus application rates on corn preceded by soybeans in two adjacent Minnesota watersheds. Source: Birr and Mulla, 2005.

fields generated a disproportionate 41 percent of phosphorus losses. Those types of disproportionalities are rarely considered in watershed remediation plans because comprehensive farm surveys are rarely conducted.

Methodological needs

Underlying the difficulties of applying new approaches to researching environmental problems in agriculture are methodological limitations. Four major needs exist for adopting a systems approach that is geographically scaleable and includes an appropriate biophysical-social science balance: geospatial data, sensor networks, modeling, and long-term experiments and assessments.

Geospatial data coverage

The ability to understand and predict environmental phenomena at landscape and regional scales requires information about the landscape and region under study. Those data must be appropriate to the questions addressed and sufficiently resolved to allow for necessary extrapolation.

Data needs are to some extent discipline-specific, but a systems approach demands that data be available for a variety of questions: Land use, terrain, climatic, soil, and crop management attributes, among others. Carbon sequestration efforts can use this information to construct spatially explicit erosivity, aggregate stability, and decomposition indices. Wildlife conservation efforts can use this information to construct spatially explicit percent-cover habitat edge indices and fragmentation corridors. Twedt and Uihlein (2005), for example, developed a spatially explicit GIS model to guide landscape-level reforestation in the Lower Mississippi Alluvial Valley, based on habitat objectives for forest interior songbirds. Proximity to extant forest tracts, connectivity, and forest core area were important predictors in this model, such that the model's accuracy depended directly upon the availability of accurate land cover data.

Hydrologic transport models illustrate the level of detail required. As noted earlier, small areas of the landscape can contribute disproportionately large amounts of sediment and phosphorus to surface waters. These small areas are often referred to as critical areas. Critical areas are characterized as hydrologically active and connected to surface waters, and they often have inappropriate management practices on steeper landscapes. This suggests that critical areas can be identified through a combination of terrain analysis, soil hydrologic pathway analysis, and farm management surveys.

The most common terrain indices for identifying critical areas are slope steepness, slope curvature, the compound terrain index (Moore et al., 1991; Gallant and Wilson, 2000), and the stream power index (Moore et al., 1993). Each terrain index is indirectly related to landscape hydrology. Slope steepness affects runoff and surface flow velocity. Slope curvature affects convergence or divergence of runoff. The compound terrain index indicates the potential for accumulation of soil water and locations where ponding may occur. The stream power index indicates the erosive power of surface runoff.

The accuracy of terrain indices is dependent upon several factors, including (1) the sampling location and density of elevation data, as well as the techniques used to collect the data; (2) the horizontal resolution and vertical precision used to represent elevation; (3) the algorithms used to calculate terrain attributes; and (4) the topography of the landscape (Theobald, 1989; Chang and Tsai, 1991; Florinksy, 1998). In general, higher resolution elevation data, such as the data available from laser imaging detection and ranging (LIDAR) systems, allow for improved identification of critical areas compared with the lower resolution elevation data typically provided by the U.S. Geological Survey.

To date there has been little research on combining terrain index modeling with soil hydrologic modeling. Surface runoff is a function of both terrain and soil attributes, but only terrain attributes are considered in indices such as the compound terrain index and stream power index. More research is also needed on methods for combining farm management practices and terrain indices to obtain better indices of landscape hydrology.

Sensor networks

Huge amounts of data are gathered on agricultural land each year, primarily by producers. Those data are primarily yield and soil tests. Few of these data are stored in a way that can be accessed to examine spatial or temporal trends.

On more highly engineered landscapes, it will be more feasible to design and implement sensor networks to monitor and facilitate management of an environmental resource.

Investment in a sensor network will be most warranted where certain criteria are met. First, there must be an achievable management objective that requires sensor data. Second, the sensors need to be simple and sufficiently robust for routine remote field deployment. Third, assuming that the management issue is at a regional or basin scale, the sensors need to provide data that can be used in a model to provide an integrated assessment to guide real-time or near-real-time management options.

There are few places in the United States where this sort of monitoring and management occur. One example is located in the San Joaquin River drainage in California (www.sjd.water.ca.gov/waterquality/realtime/). This effort is motivated by concern for anadromous fish populations and movement of selenium from agricultural land and wetlands into coastal waters. The main objective of the project is to facilitate control and timing of wetland and agricultural drainage to coincide with periods when dilution of flow is sufficient to meet salinity objectives. Salinity is measured at numerous points in the watershed by electrical conductivity probes. Those probes are relatively simple and robust, and they provide real-time data on downstream salinity and on salinity from agricultural land and wetlands in the San Joaquin River drainage. By increasing the frequency of meeting downstream salinity objectives, the project seeks to reduce the number and/or magnitude of releases of high quality water made specifically to meet downstream salinity. The high quality water saved can be used later to increase streamflow in the basin during critical periods for anadromous fish restoration efforts.

Modeling

"All models are wrong, some are useful." This well-known adage, attributed to physicist George Box, may be more relevant in the realm of integrated socioeconomic and biophysical models than most. To provide answers to questions related to agroenvironmental policy, we must develop integrated landscape-scale or watershed models that capture complex interactions among land use, conservation practices, land characteristics, and ambient water quality. The models must also couple those factors with land use choice models that incorporate economic and social drivers of choices. A number of studies have contributed to progress in this arena (e.g., Khanna et al., 2003; Beaulieu et al., 1998), but significant challenges remain. The overall need is for models that adequately integrate the biophysical and socioeconomic dimensions of important environmental issues. Five specific needs include the following:

1. Scale compatibility.
2. Adequate data coverage for model estimation and calibration.
3. Complete model coverage of policy-relevant choices.
4. Treatment of uncertainty within and across models.
5. Development and implementation of optimization algorithms using models.

Scale compatibility. Ideally, the appropriate choice for the spatial and temporal scale of a study depends upon the questions addressed. For example, an analysis designed to identify watersheds that would be best targeted for conservation funding, such as the CSP, might reasonably be based on aggregate units. A reasonable basis for selecting a watershed for inclusion in the program might be that average environmental benefits from conservation practice adoption are higher than in other watersheds and average costs are similar or lower than in other watersheds. Once targeted for funds, however, to identify the locations that would yield the most environmental gain for the funds allocated requires a much more spatially detailed modeling paradigm—one based on data from fields or farms.

Often, the choice of scale at which modeling takes place is limited by the availability of appropriate data and the capacity of computing technology. Riffell and Burger (2006), for example, used NRI data to evaluate CRP effects on northern bobwhite and grassland bird populations nationally. NRI data provide spatially explicit, point-level information on land use, but CRP practice data is low resolution, broadly documented as grass-legume or trees, with no information on year of enrollment, specific CRP practice, or patch size or shape. In contrast, USDA's Common Land Unit (CLU) polygon data, which contain information about enrollment year, conservation practice, and field boundaries, allow detection of more specific responses to CRP vegetation type, age of planting, and landscape configuration when coupled with National Land Cover Data. CLU data were available for only a three-state subset of the bobwhite range, however (Riffell and Burger 2006).

Researchers must also trade off the increased detail that comes from working at fine scales (plots or fields, hours or days) with the ability to handle very large data sets (e.g., the number of fields in the Corn Belt or the number of hours in a decade). Even in these cases however, a fairly aggregate scale may still generate conclusions that are valuable to the decision-maker, although it will be important to recognize the limitations of the analysis.

Another practical limitation comes in the form of the typical economic and biophysical models themselves. Landscape-based biophysical models are generally delineated along natural geographic lines, such as a watershed. Economic and social models are often fit to data that is drawn from political boundaries, such as counties or states. An economic model that predicts changes in average cropping practices in a county in response to a price increase is incompatible with a biophysical model based on a watershed. While obvious, this is not a trivial challenge to integration in many cases. On a related note, integrated modeling systems that combine overly simplistic or spatially aggregated models in one component (say, the economic model) with highly detailed and/or spatially disaggregated models in the other area (the biophysical model) will, at best, limit understanding of those systems and, at worst, lead to incorrect conclusions and possibly threaten the value of policy analysis.

Adequate data coverage for model estimation and calibration. A second data challenge facing researchers interested in developing or applying models at the landscape scale is the suitability or appropriateness of adequate data that is consistently collected and available across the entire landscape of interest and available at the appropriate scale. For example, in work reported by Kling et al. (2006), the SWAT model was linked with economic models to predict the costs of adopting a broad set of conservation practices in the Upper Mississippi River Basin and the in-stream water quality effects

(reductions in sediment, nitrogen, and phosphorus) from doing so.

While a valuable start on this important problem area, a number of shortcomings related to data availability and quality limit the model's usefulness. First, limited consistent and long-term measurement data on actual in-stream water quality across the basin makes calibration of the parameters in the SWAT model challenging. While data sources exist to accomplish calibration, there is little doubt that more consistent, thorough, and well-documented data would improve the accuracy and believability of model runs.

A second example from this same modeling effort concerned the use and distribution of manure and fertilizer applications on cropped fields in the region. While statewide fertilizer sales are reported, a significant amount of guesswork concerning where this purchased fertilizer is spread in conjunction with where the manure from hog and chicken facilities in the region is applied. While a number of self-reported surveys have been administered and reports from that data are available, there are discrepancies and inconsistencies that make this information hard to rely upon without adjustment to the statewide information.

An example from the economics portion of the integrated model is the availability of cost information on the construction and adoption of conservation practices across multiple states. While Natural Resources Conservation Service (NRCS) offices in each state may collect and report county-level cost-share information for a variety of state and federal conservation programs, those data are not uniformly reported and not available consistently or in a central location. Thus, key information necessary for model building and model calibration are often lacking. Without consistent, long-term data collection to fit and calibrate landscape-level models, the accuracy and usefulness of the models cannot be significantly improved.

Complete model coverage of policy-relevant choices. While data sources may represent a major stumbling point in developing and implementing models, another is a lack in the available modeling capacity itself. Again, drawing from the Upper Mississippi River Basin SWAT-economic model, a major management goal is to reduce sedimentation and movements of phosphorous and other nutrients from farm fields into waterways of the watershed. The SWAT model provides a critically valuable tool on the biophysical side, but does not yet contain methods to site perennial buffers or wetlands within the watershed. Thus, conservation practices, such as reduced tillage, grassed waterways, and land retirement, can all be handled in various modeling scenarios, but a combination of practices with carefully targeted siting of wetlands and buffers cannot. This limits the ability of the modeling paradigm to evaluate many of the policy options that would ideally be considered in planning watershed management for TMDLs in the region or to meet other water quality goals.

In considering the costs of adopting alternative conservation practices, it also is clear that the costs of accepting increased risk in profits and the time costs of farmers belong in the computation of the costs of adopting a particular conservation practice (such as moving from fall to spring fertilizer application). While economists recognize these as appropriate costs and a literature concerning those costs has developed, much is still not understood about the magnitude of the costs and how they vary across farming operations, crops, and locations. Particularly challenging is the cost of nutrient management, which often entails significant learning and time on the part of the farm manager; moreover, there remains a robust debate among agronomists over the short- and long-term yield consequences of reduced nutrient application levels.

Treatment of uncertainty within and across models. A major challenge within the entire scientific community is in reporting and representing uncertainty associated with scientific findings in a way that presents a fair representation of its sources and magnitudes without completely undermining the value of the message. Many of the same issues apply to the case of integrating biophysical and socioeconomic modeling, but here the challenges may be even greater, for two reasons. First, the nature of these models suggest that a great many variables, data sources, and individual models will have to be combined to construct the modeling system and run appropriate scenarios. Each of the variables and models will have various sources of uncertainty associated with them

that will, in many cases, be unknown or difficult to quantify.

Second, even if all sources of individual uncertainty could be quantified, many of the sources of uncertainty across models may be correlated. If so, simple addition of weighted standard errors will not yield correct measures of the aggregate variability. Instead, it will be necessary to correct for correlations that can raise or lower the aggregate uncertainty, depending upon the sign of correlation.

Development and implementation of optimization algorithms using models. The challenges associated with integrated modeling do not end once the model is developed. From an economic standpoint, the optimal combination of conservation practices and their location is the spatial configuration of practices that achieves the water quality target at the least possible cost. Unfortunately, as others have recognized (Lintner and Weersink 1999; Khanna et al., 2003), studying the least-cost solution in this context is challenging.

First, the optimal combination may vary dramatically with the chosen level of water quality. That is, the set of conservation practices that will appear in a least-cost solution to achieve a relatively small water quality improvement may differ vastly from the set of practices that appear when a high level of water quality is the objective.

Second, because of the unique nature of the biophysical relationship between conservation practices and resulting water quality levels, the impact of a farmer's conservation practice decision on his field depends also upon the choices of all others' conservation practices, cropping systems, and related decisions in the watershed.

Third, there are multiple abatement possibilities for each field in that many different conservation practices could be implemented. This means that identifying the least-cost solution requires comparing a large number of possible land use scenarios. Specifically, if there are n conservation practices possible for F fields, there will be a total of nF possible configurations. In a watershed with hundreds of fields and eight to ten conservation practices, this comparison quickly becomes unwieldy. An important bright spot in this area is the adaptation of genetic algorithms to watershed models. Those methods were developed for use in general nonlinear and correlated systems, and recent experience suggests they may be useful for approximating the optimal solution to conservation practice placement in a watershed.

Long-term studies and outcome assessment

Comprehensive records are kept on a county basis for all practices installed under USDA programs through the NRCS Performance Results System (PRS). This system tracks most conservation activity on private land in the United States. For most land treatment, the system keeps track of acres treated. For certain practices, such as comprehensive nutrient management planning (CNMP), PRS keeps track of the number of contracts written and the number of acres applied. While the system forms a reasonable basis for estimating cumulative effects of conservation practices, the measurements in PRS are only a first step in understanding the effects of conservation practices. Other measurements are needed to convert PRS data into benefits of conservation practices.

Data on per acre savings in soil, nitrate, and carbon loss and per acre wildlife benefits are also needed. Those estimates are available from only a few studies or modeling efforts. More measurements are needed. Consider that despite billions of dollars invested in the CRP no study has systematically evaluated wildlife response to the CRP from its inception. Presumptive wildlife benefits are inferred from the cumulative weight of evidence from a myriad of site-specific studies (Haufler, 2005). Only recently have wildlife benefits been systematically evaluated on a random sample of contracts for a specific CRP practice (CP33) (Burger et al., 2006).

What is also needed is an understanding of the condition of the practices. Dillaha et al. (1989) studied existing grass filter strips on 18 farms in Virginia and found them to be extremely variable in their effectiveness at removing sediment. Most grass filter strips in hilly areas were ineffective because runoff usually crossed the strip as concentrated flow. In flatter regions, grass filter strips were more effective because slopes were more uniform and more runoff entered the strip as shallow flow. Several one- to three-year-old vegetative filter strips were observed to have trapped so much sediment that they produced more sedi-

ment than adjacent upland fields. In those cases, runoff flowed parallel to the vegetative filter strip until it reached a low point, where it crossed the vegetative filter strip as concentrated flow. The vegetative filter strips clearly needed maintenance to regain their sediment-trapping abilities.

These experiences point out the need for tracking not only the application of practices, but the maintenance of practices as well. Annual inspections of practices, such as structural measures (e.g., terraces and grass waterways) and permanent vegetative cover should be possible on all farms receiving cost-share or technical assistance. Other methods, including remote sensing techniques, are needed to understand whether practices, such as residue management and cover crops, are being applied properly. Unfortunately, many of the most important conservation practices, such as nutrient management and pest management, have many aspects that cannot be verified or inspected visually.

Summary and conclusions

We conclude the following:

1. Traditional field-scale approaches to environmental research and management have been highly effective in identifying specific field-edge solutions for individual environmental problems. They have been less successful in identifying solutions that are robust across multiple spatial scales, that are socially acceptable, and that meet multiple goals.

2. Needed now are approaches that are systems-oriented (in order to understand direct and indirect outcomes and to balance multiple aims), that are geographically scaleable (to allow effective implementation at watershed and regional scales), that incorporate socioeconomic factors (to allow for the inclusion of human behavior and effective incentives), and that provide for long-term outcome monitoring (to allow solutions to be assessed over time periods that include diverse climatic and other environmental events).

3. To implement those approaches will require resources to provide, in particular, (a) adequate geospatial data coverage, (b) development and deployment of new sensors that

can monitor key environmental attributes over appropriate geographic areas, (c) new modeling approaches that provide for the effective integration of existing biophysical and socioeconomic models, and (d) long-term experiments and outcome assessments.

4. New research, systems-oriented at appropriate geographic and temporal scales, could provide the knowledge needed to develop and implement effective policies for delivering the ecological services that agriculture can provide.

References

Aneja, V. P., B. Y. Wang, D. Q. Tong, H. Kimball, and J. Steger. 2006. Characterization of major chemical components of fine particulate matter in North Carolina. *Journal of the Air & Waste Management Association* 56: 1099-1107.

Beaulieu, J., D. Bennett, S. Kraft, and R. Segupta. 1998. Ecological-economic modeling on a watershed basis: A case study of the Cache River of Southern Illinois. *American Agricultural Economics Association*, Salt Lake City, Utah.

Birr, A. S., and D. J. Mulla. 2005. Paired watershed studies for nutrient reductions in the Minnesota River Basin. http://www.usawaterquality.org/integrated/MNRiver.html.

Braden, J. B., G. V. Johnson, A. Bouzaher, and C. Miltz. 1989. Optimal spatial management of agricultural pollutions. *American Journal of Agricultural Economics* 78: 961-971.

Brinson, M. M., R. D. Rheinhardt, F. R. Hauer, L. C. Lee, W. L. Nutter, R. D. Smith, and D. F. Whigham. 1995. A *guidebook for application of hydrogeomorphic assessments to riverine wetlands*, WRP-DE-11. U.S. Army Engineer Waterways Experiment Station, Vicksburg, Mississippi.

Brinson, M. M., and R. D. Rheinhardt. 1996. The role of reference wetlands in functional assessment and mitigation. *Ecological Applications* 6: 69-76.

Burger, L. W. 2005. The Conservation Reserve Program in the Southeast: issues affecting wildlife habitat value. In J. B. Haufler, editor. *Fish and Wildlife Benefits of Farm Bill Conservation Programs: 2000–2005 Update*. Technical Review 05-2. The Wildlife Society Bethesda, Maryland. Pp 63–92.

Burger, L. W. Jr., D. McKenzie, R. Thackston, and S. DeMaso. 2006. The role of farm policy in achieving large-scale conservation: Bobwhite and buffers. *Wildlife Society Bulletin*. 34: 986-993.

Calhoun, F. G., B. K. Slater, and D. B. Baker. 2002. Soils, water quality, and watershed size: interactions in the Maumee and Sandusky River basins of northwestern Ohio. *Journal of Environmental Quality* 31: 47-53.

Carpentier, C. L., D. Bosch, and S. S. Batie. 1998. Using spatial information to reduce costs of controlling agricultural nonpoint source pollution. *Agricultural and Resources Review* 27: 72-84.

Chalfoun, A. D., F. R. Thompson III, and M. J. Ratnaswamy. 2002. Nest predators and fragmentation: a review and meta-analysis. *Conservation Biology* 16:306-318

Chang, K. T., and B. W. Tsai. 1991. The effect of DEM resolution on slope and aspect mapping. *Cart. Geog. Info. Sys.* 18: 69-77.

Clark, W. R. and K. F. Reeder. 2005. Continuous enrollment Conservation Reserve Program: factors influencing the value of agricultural buffers to wildlife conservation. In J. B. Haufler, editor. *Fish and Wildlife Benefits of Farm Bill Conservation Programs: 2000–2005 Update*. Technical Review 05-2. The Wildlife Society, Bethesda, Maryland. Pp. 93 – 114.

Crumpton, W. G. 2000. Using wetlands for water quality improvement in agricultural watersheds: The importance of a watershed scale approach. *Water Science and Technology* 44: 559-564.

Crumpton, W. G. 2005. Potential of wetlands to reduce agricultural nutrient export to water resources in the Corn Belt. *Gulf Hypoxia and Local Water Quality Concerns Workshop*. Iowa State University. Pp. 35-44.

Davie, D. K., and C. L. Lant. 1994. The effect of CRP enrollment on sediment in two Illinois streams. *Journal of Soil and Water Conservation* 49: 407-412.

Dillaha, T. A., J. H. Sherrard, and D. Lee. 1989. Long-term effectiveness of vegetative filter strips. *Water, Environment and Technology* 1: 419-421.

Elmi, A. A., C. Madramootoo, M. Egeh, and C. Hamel. 2004. Water and fertilizer nitrogen management to minimize nitrate pollution from a cropped soil in southwestern Quebec, Canada. *Water, Air, and Soil Pollution* 151: 117-134.

Farrand, D. T. and M. R. Ryan. 2005. Impact of the Conservation Reserve Program on wildlife conservation in the Midwest. In J. B. Haufler, editor. *Fish and Wildlife Benefits of Farm Bill Conservation Programs: 2000–2005 Update.*

Technical Review 05-2. The Wildlife Society Bethesda, Maryland. Pp. 41–62.

Feng, H., L. Kurkalova, C. Kling, and P. Gassman. 2006. Environmental conservation in agriculture: Land retirement vs. changing practices on working land. *Journal of Environmental Economics and Management* 52: 600-614.

Feng, H., L. Kurkalova, C. Kling, and P. W. Gassman. 2007. Transfers and environmental co-benefits of carbon sequestration in agricultural land in the Upper Mississippi River Basin. *Climatic Change* 80:91-107.

Ferriter, A. 1997. Brazilian pepper management plan for Florida. Florida Exotic Plant Council.

Findlay, S. E., E. Kiviat, W. C. Nieder, and E. A. Blair. 2002. Functional assessment of a feference wetland set as a tool for science, management, and restoration. *Aquatic Sciences* 64: 107-117.

Florinsky, I. V. 1998. Accuracy of local topographic variables derived from digital elevation models. *International Journal of Geographical Information Science* 12: 47-61.

Gallant, J. C., and J. P. Wilson. 2000. Primary topographic attributes in J. P. Wilson and J. C. Gallant, eds. *Terrain Analysis: Principals and Applications*. John Wiley and Sons, Inc., New York, New York.

Gburek, W. J., C. C. Drungil, M. S. Srinivasan, B. A. Needleman, and D. E. Woodward. 2002. Variable source area controls on phosphorus transport: bridging the gap between research and design. *Journal of Soil and Water Conservation* 57: 534-543.

Haufler, J. B., editor. 2005. *Fish and Wildlife Benefits of Farm Bill Conservation Programs: 2000–2005 Update.* Technical Review 05-02. The Wildlife Society, Bethesda, Maryland.

Horn, D. J., M. L. Phillips, R. R. Koford, W. R. Clark, M. A. Sovada, and R. J. Greenwood. 2005. Landscape composition, patch size, and distance to edges: Interactions affecting duck reproductive success. *Ecological Applications* 15: 1367-1376.

Intergovernmental Panel on Climate Change. 2002. *IPCC Special Report on Land Use, Land-Use Change and Forestry*. Cambridge University Press, Cambridge, United Kingdom.

Jaynes, D. B., D. L. Dinnes, D. W. Meek, D. L. Karlen, C. A. Cambardella, and T. S. Colvin. 2004. Using the late spring nitrate test to reduce nitrate loss within a watershed. *Journal of Environmental Quality* 33: 669-677.

Johansson, R. C., P. H. Gowda, D. J. Mulla, and B. J. Dalzell. 2004. Metamodeling phosphorus best management practices for policy use: A frontier approach. *American Journal of Agricultural Economics* 30: 63-74.

Johnson, D. H. 2005. Grassland bird use of Conservation Reserve Program fields in the Great Plains. In J. B. Haufler, editor. *Fish and Wildlife Benefits of Farm Bill Conservation Programs: 2000–2005 Update*. Technical Review

05-02. The Wildlife Society, Bethesda, Maryland. Pp. 17–32.

Khanna, M., W. Yang, R. Farnsworth, and H. Onal. 2003. Cost-effective targeting of land retirement to improve water quality with endogenous sediment deposition coefficients. *American Journal of Agricultural Economics* 85: 538-553.

Kling, C., S. Secchi, M. Jha, H. Feng, P. W. Gassman, and L. A. Kurkalova. 2006. Upper Mississippi River Basin modeling system part 3: Conservation practice scenario results. In V. P. Singh and Y. J. Xu, editors, *Coastal Hydrology and Processes*. Water Resources Publications, Highland Ranch, Colorado.

Kovacic, D. A., M. B. David, L. E. Gentry, K. M. Starks, and R. A. Cooke. 2000. Effectiveness of constructed wetlands in reducing nitrogen and phosphorus export from agricultural tile drainage. *Journal of Environmental Quality* 29: 1262-1274.

Kurkalova, L., C. Kling, and J. Zhao. 2006. Green subsidies in agriculture: estimating the adoption costs of conservation tillage from observed behavior. *Canadian Journal of Agricultural Economics* 54: 247-267.

Landis, D. A., F. D. Menalled, A. C. Costamagna, and T. K. Wilkinson. 2005. Manipulating plant resources to enhance beneficial arthropods in agricultural landscapes. *Weed Science* 53: 902-908.

Lintner, A. M., and A. Weersink. 1999. Endogenous transport coefficients. *Environmental and Resource Economics* 14: 269-296.

Lowrance, R. R., R. L. Todd, and L. E. Asmussen. 1984. Nutrient cycling in an agricultural watershed. II. Streamflow and artificial drainage. *Journal of Environmental Quality* 13: 27-32.

Lowrance, R. 1998. Riparian forest ecosystems as filters for nonpoint-source pollution. Pages 113-141 in M. L. Pace and P. M. Groffman, eds. *Successes, Limitations, and Frontiers in Ecosystem Science*. Springer Verlag, New York, New York.

Mausbach, M. J., and A. R. Dedrick. 2004. The length we go: measuring environmental benefits of conservation practices. *Journal of Soil and Water Conservation* 59: 96-103.

Moore, I. D., R. B. Grayson, and A. R. Ladson. 1991. Digital terrain modeling: a review of hydrological, geomorphological and biological applications. *Hydrology Proceedings* 5: 3-30.

Moore, I. D., P. E. Gessler, G. A. Nielsen, and G. A. Peterson. 1993. Soil attribute prediction using terrain analysis. *Soil Science Society of America Journal* 57: 443-452.

National Research Council. 2000. Clean coastal waters: Understanding and reducing the effects of nutrient pollution. National Research Council, National Academy Press, Washington, D.C.

National Research Council. 2003. Frontiers in agricultural research. Food, health, environment, and communities. National Academy Press, Washington, D.C, USA.

Petrolia, D., P. H. Gowda, and D. J. Mulla. 2005. Targeting agricultural drainage to reduce nitrogen losses in a Minnesota watershed. Staff Paper Series P-05-2. Department of Applied Economics, University of Minnesota. http://agecon.lib.umn.edu/mn/p05-02.pdf.

Phillips, M. L., W. R. Clark, S. M. Nusser, M. A. Sovada, and R. J. Greenwood. 2004. Analysis of predator movement in prairie landscapes with contrasting grassland composition. *Journal of Mammalogy* 85:187 – 195.

Phillips, M. L., W. R. Clark, M. A Sovada, D. J. Horn, R. R. Koford, and R. J. Greenwood. 2003. Predator selection of prairie landscape features and its relation to duck nest success. *Journal of Wildlife Management* 67:104 – 114.

Qiu, Z., and T. Prato. 1999. Accounting for spatial characteristics of watersheds in evaluating water pollution abatement policies. *Journal of Agricultural and Applied Economics* 31: 161-175.

Randall, G. W., and J. E. Sawyer. 2005. Nitrogen application timing, forms, and additives. Gulf Hypoxia and Local Water Quality Concerns Workshop. Iowa State University, Ames, Iowa. Pp. 73-84.

Reynolds, R. E. 2005. The Conservation Reserve Program and duck production in the U.S. Prairie Pothole Region. In J. B. Haufler, editor, Fish and Wildlife Benefits of Farm Bill Conservation Programs: 2000–2005 Update. Technical Review 05-2. The Wildlife Society Bethesda, Maryland. Pp. 17–32.

Ribaudo, M., R. Heimlich, R. Claasen, and M. Peters. 2001. Least-cost management of nonpoint source pollution: source reduction versus interception strategies for controlling nitrogen loss in the Mississippi basin. *Ecological Economics* 37: 183-197.

Richards, R. P., and D. B. Baker. 2002. Trends in water quality in LEASQ rivers and streams (Northwestern Ohio), 1975-1995. *Journal of Environmental Quality* 31: 90-96.

Richardson, C. J., and S. S. Qian. 1999. Long-term phosphorus assimilative capacity in freshwater wetlands: A new paradigm for sustaining ecosystem structure and function. *Environmental Science & Technology* 33: 1545-1551.

Riffell, S. K. and L. W. Burger. 2006. Estimating Wildlife Response to the Conservation Reserve Program: Bobwhite and Grassland Birds. Final report for FSA-R-28-04DC Estimating Wildlife Response to the Conservation Reserve Program, Department of Agriculture, Farm Service Agency, Acquisition Management Branch, Special Projects Section. http://www.fsa.usda.gov/Internet/FSA_File/quail_study.pdf (accessed 4/19/07).

Robertson, G. P. 2004. Abatement of nitrous oxide, methane, and the other non-CO2 greenhouse gases: The need for a systems approach. In C. B. Field and M. R. Raupach, editors, The Global Carbon Cycle. Island Press, Washington, D.C. Pp. 493-506.

Robertson, G. P., and S. M. Swinton. 2005. Reconciling agricultural productivity and environmental integrity: A grand challenge for agriculture. *Frontiers in Ecology and the Environment* 3: 38-46.

Robertson, G. P., J. C. Broome, E. A. Chornesky, J. R. Frankenberger, P. Johnson, M. Lipson, J. A. Miranowski, E. D. Owens, D. Pimentel, and L. A. Thrupp. 2004. Rethinking the vision for environmental research in U.S. agriculture. *BioScience* 54: 61-65.

Russell, C. S., and J. F. Shogren. 1993. Theory, modeling and experience in the management of nonpoint-source pollution. Kluwer Academic Publishers.

Ryan, M. R. 2000. Impact of the Conservation Reserve Program on wildlife conservation in the Midwest. In W. L. Hohman and D. J. Halloum, editors, A Comprehensive Review of Farm Bill Contributions to Wildlife Conservation, 1985–2000. Technical Report USDA/NRCS/WHMI-2000. U.S. Department of Agriculture, Forest Service, Washington, D.C. Pp. 45–54.

Soil and Water Conservation Society. 2006. Final report from the Blue Ribbon Panel conducting an external review of the U.S. Department of Agriculture Conservation Effects Assessment Project. Ankeny, Iowa.

Stephens, S. E., D. N. Koonsa, J. J. Rotella, and D. W. Willey. 2003. Effects of habitat fragmentation on avian nesting success: a review of the evidence at multiple spatial scales. *Biological Conservation* 115: 101–110.

Sunding, D., and D. Zilberman. 2000. The agricultural innovation process: research and technology adoption in a changing agricultural sector. In B. Gardner and G. Rausser, editors, Handbook of Agricultural Economics. Elsevier Science Publishers.

Swinton, S. M., F. Lupi, G. P. Robertson, and D. A. Landis. 2006. Ecosystem Services from Agriculture: Looking Beyond the Usual Suspects. *American Journal of Agricultural Economics* 88:1160-1166

Tan, C. S., C. F. Drury, W. D. Reynolds, P. H. Groenvelt, and H. Dalfar. 2002. Water and nitrate loss through tiles under a clay loam soil in Ontario after 42 years of consistent fertilization and crop rotation. *Agriculture, Ecosystems & Environment* 93: 121-130.

Theobald, D. M. 1989. Accuracy and bias issues in surface representation. In M. Goodchild and S. Gopol, editors, Accuracy of Spatial Databases. Taylor and Francis, Bristol, Pennsylvania. Pp. 99-106.

Thompson, F. R. III, T. M. Donovan, R. M. DeGraaf, J. Faaborg, and S. Robinson. 2002. A multi-scale perspective on the effects of forest fragmentation on birds in eastern forests. *Studies in Avian Biology* 25:8 – 19.

Twedt, D. J. and W. B. Uihlein III. 2005. Landscape level reforestation priorities for forest breeding landbirds in the Mississippi Alluvial Valley. In L. H. Fredrickson, S. A. King, and R. M. Kaminski, editors, Ecology and Management of Bottomland Hardwood Systems: The State of our Understanding. Gaylord Memorial Laboratory Special Publication No. 10, Puxico.University of Missouri-Columbia. Pp. 321-340.

Udawatta, R. P., J. J. Krstansky, G. S. Henderson, and H. E. Garrett. 2002. Agroforestry practices, runoff, and nutrient loss: A paired watershed comparison. *Journal of Environmental Quality* 31: 1214-1225.

Vellidis, G., R. Lowrance, P. Gay, R. W. Hill, and R. K. Hubbard. 2003. Nutrient transport in a restored riparian wetland. *Journal of Environmental Quality* 32: 711-726.

Ward, A. D., J. L. Hatfield, J. A. Lamb, E. E. Alberts, T. J. Logan, and J. L. Anderson. 1994. The management systems evaluation areas program: tillage and water quality research. *Soil Tillage and Research* 30: 49-74.

Wilson, A. T. 1978. The explosion of pioneer agriculture: contribution to the global CO_2 increase. *Nature* 273: 40-41.

Zedler, J. B., and S. Kercher. 2004. Causes and consequences of invasive plants in wetlands: Opportunities, opportunists, and outcomes. *Critical Reviews in Plant Science* 23: 431-452.

Roundtable:
Methods for environmental management research

At the outset of the roundtable, participants reiterated two important goals of the U.S. Department of Agriculture's Conservation Effects Assessment Project: (1) Estimate the effects of conservation practices and (2) expand those estimates for the entire nation. Discussion then focused on several important talking points:

- Do we need standardized research methods? In that practical given different ecoregions, microclimates, geomorphology, and so forth?

- Do we have a consensus on the "state of the art" in terms of methods and approaches? Do we know what everyone is doing?

- Are we gathering the right types of data? Are those types of data what the modelers will need?

- Are our methods suitably precise?

- Do we agree on what our practices are supposed to be doing, for example, reducing sediment, enhancing wildlife, processing nutrients?

- Using our methods, can we detect confounding effects or conflicting effects? For example, one conflicting effect: Winter cover reduces sediment, but may increase organic loading, which may reduce dissolved oxygen. One confounding effect: Reduced sediment increases light, which increases chlorophyll a, but reducing sediment also reduces phosphorus, which reduces chlorophyll a.

Roundtable participants then identified the six most important "next steps" necessary to strengthen the science associated with environmental management research methods or approaches:

1. Better understanding of current model limitations and development of models that account for larger scale processes (stream channel processes, reaction kinetics). If important processes are not adequately represented, calibration may match results for the wrong reasons, and extrapolation to changed conditions will not reflect real results.

2. Long-term (30 or more years) geospatial datasets suitable for calibrating and testing the next generation of models (long-term geospatial data at different scales, including waterways). Get major players (U.S. Geological Survey, U.S. Environmental Protection Agency, U.S. Department of Agriculture,…) to work together to populate and maintain a data network.

3. Standardize methodologies for data collection, sampling protocols, and analytical procedures. Prepare quality control and assurance procedures.

4. Standardization and/or development of a methodology to declare "significant differences or effects." Techniques for analyzing uncontrollable, unreplicable situations. Error estimates, variance components, meta-analysis, time series. What are the acceptable measures to declare that "we did or did not make a difference"?

5. Identify specific experiments that need to be done across all watershed studies as work moves from the field level to the watershed level. Provide the infrastructure and support to conduct experiments of processes at multiple spatial and temporal scales (like long-term ecological research, multiple paired watersheds, and so forth).

6. Develop tools to tailor adaptive management to an agricultural watershed context.

Coping with the data dilemma

F.J. Pierce
P. Nowak
P.E. Cabot

Scientists addressing larger-scale problems (e.g., continental drift, climate change) have never found it easy to have their ideas accepted ... because the scientific method works best when the object of study is well defined, can be isolated from extraneous influences, and small changes in the conditions can be made at will. Then the classical scientific method of hypothesis testing by experiment is relatively easy. Unfortunately, many of the urgent problems that we face in the world are large-scale problems to do with land and resource utilization which ... are not especially amenable to this approach. (Grace et al., 1997)

The Conservation Effects Assessment Project (CEAP) began in 2003 to quantify the environmental benefits of conservation practices funded by U.S. Department of Agriculture (USDA) conservation programs (Soil and Water Conservation Society, 2006). Convened not long after the program began, a blue ribbon panel reviewed CEAP and recommended it change direction immediately to insure that the project improves future conservation efforts (Soil and Water Conservation Society, 2006). The panel's report included a specific recommendation that "simulations and extrapolations cannot – and must not – substitute for on-the-ground monitoring and inventory systems designed to determine if anticipated conservation and environmental benefits are being achieved" (Soil and Water Conservation Society, 2006).

This recommendation parallels an earlier and critical U.S. Geological Survey (USGS) report on the lack of improvement in ecological condition and water quality in the Chesapeake Bay, despite 20 years of restoration efforts (Phillips, 2005). The SWCS panel recommendation was intended to insure that such a dismal result does not ultimately characterize the CEAP initiative. The USGS report called for more effective integration of science with implementation of conservation and restoration activities that "will help promote more of an adaptive management approach by having more integrated measurements of the ecosystem response to management activities, climate variability, and to use forecasting to help revise development of new management approaches" (Phillips, 2005). Understanding the outcome of large, complex human and ecological interactions is not easy, and any assessment of this situation must reflect this fact.

In the foregoing chapter of this book, Robertson and colleagues delivered a similar message, suggesting that extensive monitoring and modeling both are needed at multiple scales to shift effectively from field-scale to landscape approaches to managing agricultural watersheds. Groffman and colleagues, earlier in this book, suggested that such a broad-scale effort will require improvements in both monitoring and information technologies, while Robertson and his coauthors suggested that four major resource requirements are needed, including adequate geospatial data coverage, development and deployment of new sensors for monitoring key environmental attributes over appropriate geographic areas, new modeling approaches that provide for effective integration of existing biophysical attributes and socioeconomic models, and long-term experiments and outcome assessments. More monitoring and more modeling across multiple scales designed to integrate biophysical dimensions with social dimensions of landscape management connote significant increases in data collection, management, and analysis. It is highly unlikely that our current data systems can meet those requirements. This, then, is the data dilemma associated with managing agricultural watersheds. Coping with this data dilemma means successfully exploring alternative approaches through a careful, strategic assessment of the current situation and an open, innovative perspective on how to best address this dilemma.

The data dilemma for managing agricultural landscapes is not new, unexpected, or unique. Ward et al. (1986) used the "data rich, information poor" (DRIP) syndrome to describe the state of water quality monitoring in the mid-1980s. Natural Resource Conservation Service (NRCS) leaders admitted in 1995 that that agency suffered from the DRIP syndrome because it had significant data in largely unusable forms (Johannsen et al., 1995). The DRIP syndrome is also known to hinder the health care and insurance industries. Goodwin (1996) suggested that the DRIP syndrome was paralyzing the performance improvement efforts of many healthcare organizations due to the use of an abundance of indicators and the predominant use of a retrospective medical record review to collect data.

The "State of the Nation's Ecosystems" report, first released in 2002, had the goal of identifying and reporting on a set of high-level indicators on the status of the nation's ecosystems (H. John Heinz III Center for Science, Economics, and the Environment, 2002.). That report, based on existing data collection efforts and data bases, found the state of knowledge on many environmental issues to be sorely lacking, so much so that it was not possible to quantify the actual condition of our nation's resources. Significant data gaps remain that preclude a thorough report of the nation's ecosystem health. Those significant data gaps are due both to a lack of data and a lack of data integration from disparate sources (multiple agencies and institutions).

The data dilemma relative to managing agricultural landscapes is similar in many ways. There are myriad data sources from disparate sources that bear on a problem organized by different rationales for data collection. There is little standardization in the types of data collected, and what data are collected vary over a range of spatial and temporal scales. Finally, the data are not integrated in systematic fashion to allow for assessment of environmental performance across space or time.

While there is no easy answer to this data dilemma, our objective is to provide insights on

how to cope effectively with the data dilemma associated with managing agricultural systems at the landscape level. Specifically, we discuss four relevant elements for any successful management strategy on agricultural landscapes: (1) Technological innovation, (2) scaling, (3) execution, and (4) profiles of data users.

Technology—where there's a will, there's a way

Even today's technology and knowledge can reduce considerably the human impact on ecosystems. They are unlikely to be deployed fully, however, until ecosystem services cease to be perceived as free and limitless, and their full value is taken into account. (Millennium Ecosystem Assessment, 2005)

In 1965 Gordon Moore, co-founder of Intel, proposed that the number of transistors placed on an integrated circuit would double every 24 months, a paradigm later known as Moore's Law of Integrated Circuits (Moore, 1965). Moore's Law is but one example of what Kurzweil (2001) refers to as technologies that follow the law of accelerating returns. That law purports that changes in technology, such as computational power and speed, are exponential.

In the two decades since passage of the 1985 farm bill that included the conservation programs evaluated by CEAP, technology for acquiring, storing, managing, and analyzing data has increased exponentially. But technological innovation has outpaced use of that innovation in conservation management. The point to be made here is that technological innovation may allow us to meet the multiscale data requirements needed for monitoring and adaptive management if we "will" to pursue it. Because change is episodic, nonlinear, and constant, the only feasible approach is to embrace technological change in our data management strategies.

A case in point involves the wireless sensor networks now being deployed in regional and on-farm applications. By integrating wireless sensor networks with the internet, it is possible to monitor ecosystem functions at the landscape level in real-time, with easy access to current and historical data via the internet. For example, an extensive wireless sensor network is installed at the University of California James San Jacinto Mountains Reserve, located on the western flank of Black Mountain in Riverside County (http://www.jamesreserve.edu/factoids.html). A click on that website to the link "Google Earth Interface to the James Reserve Observing Systems" takes you to the Google Earth web browser. There, on-line access is available to multiple microclimate and reference weather stations, video feeds from cameras situated on towers and inside bird houses, fixed and mobile sensor platforms for soils, and aquatic and above-ground measurements of a variety of ecological processes (http://www.james-reserve.edu/factoids.html). Extensive geospatial data sets are also available for the James Reserve at the main website. This example illustrates how real-time research at the watershed scale can be performed, but it also exemplifies how anyone who needs or chooses to examine this data can do so with an easy, accessible, user-friendly, and universal format (Google Earth).

There is also value in wireless sensor networks on farms where high-density data sets are generated in real-time. Pierce and Elliott (2007) developed an on-farm wireless sensor network initially intended for use in frost monitoring. Fruit growers use various technologies to raise air temperatures in orchards and vineyards when conditions portend low-temperature damage to plants. Frost protection (wind machines, irrigation, and smudge pots) is initiated on the basis of threshold temperatures. A wireless sensor network, consisting of 900 MHz, frequency-hopping, spread-spectrum radios, was deployed in a base-remote network topology on the farm. This network records air temperature every minute for the grower and farm workers operating a client-software program called AgFrostNet on laptop PCs connected to a roaming radio (Pierce and Elliott, 2007). Each user sees current and historical data trends on the computer screen and makes decisions to implement frost protection strategies based on observed, real-time data. Those data can also be made available in real-time on the internet by simply linking the laptop via an ethernet connection. Many other possibilities exist for use of wireless sensor networks on farms for monitoring and managing farm operations, such as irrigation (Pierce et al., 2006), and ecosystem services, such as water quality.

Technologies are advancing through miniaturization of electronics ("lab on a chip"), advances in processor and graphics processing units, increasing availability of open-source software, and universal internet services (e.g., Google Earth). In combination with such technologies as wireless sensor networks, it is feasible to generate and visualize high-density data sets across spatial-temporal scales. These new technologies provide opportunities to enhance management at spatial and temporal scales that would have been impossible even a few years ago. But the real lesson to be derived from this increasing pace of technological innovation lies in a strategic assessment of its potential. Not all agricultural landscapes or all processes occurring within those landscapes require the application of innovative technologies. The generation of decision rules to guide the application of innovative technologies is as critical as the technology itself. Those decision rules can be derived from either salient scientific theories or from political outcomes embedded in program guidelines. Whatever their source, the decision rules will need to address the issue of scale.

Scaling—large-scale problems may involve small-scale solutions

Local processes sometimes spread to become important regionally or globally, but ecosystem services at more aggregated scales are seldom simple summations of the services at finer scales.... Conversely, most services are delivered at the local scale, but their supply is influenced by regional or global-scale processes...our capability of predicting emergence of cross-scale effects and their impacts on ecosystem services is limited. A related problem is the mismatch between the scales at which natural and human systems organize. (Carpenter et al., 2006)

CEAP is focused on scaling up from the field to the watershed to improve ecosystem services that operate at larger scales. To some that might intuitively imply that approaches to managing agricultural landscapes in general also need to scale up. While that may be true in some cases, perhaps in cases where data are limited or there are scale mismatches (Carpenter et al., 2006), a case can be made for scaling down where large-scale problems involve small-scale solutions. A case in point is nitrogen fertilizer management in wheat (*Triticum aestivum* L.). In general, fertilizer use efficiencies in cereal grains is low, reportedly averaging about 33 percent for cereal crops (Raun and Johnson, 1999); this provides considerable potential for nitrogen losses to the environment (Galloway et al., 2003; Fenn et al., 2003). Using optical sensors to detect simultaneously the nitrogen status of wheat between Feekes growth stages 4 through 6, Raun et al. (2005) developed a system to trigger the application of nitrogen fertilizer if a deficiency is detected in real-time; their nitrogen fertilization rate is a function of in-season estimates of potential yield and the likely yield response to nitrogen fertilizer. This in-season nitrogen management method applies nitrogen fertilizer at a resolution of 1 square meter (1.1 square yards), resulting in a 15 percent increase in nitrogen use efficiency over whole-field nitrogen management methods (Raun et al., 2002, 2005). The scale of management is 1 square meter, but the ecosystem benefit is landscape scale. The point is that the principles and practices of precision agriculture are perfectly aligned with the principles of managing agricultural landscapes (Pierce and Nowak, 1999).

Other examples of landscape solutions being found at smaller scales in space or time are related to the degradation that may occur from agricultural production processes. Sharpley and colleagues (2001) found that small critical source areas in a watershed are responsible for a significant proportion of overall phosphorus losses. Shipitalo and Owens (2006) found that a large proportion of pesticide losses in agricultural landscapes occur during a few storm events across time. In similar fashion, it is well-known that a few large storm events (Edwards and Owens, 1991) cause the majority of soil erosion on agricultural landscapes. The lesson from this research is that managing agricultural landscapes to enhance ecological goods and services may require managing small portions of space, or even for short durations of time, within the landscape.

Execution—the key

Never is execution more important than when innovation is at the heart of a strategy. That is because innovation always involves treading into uncertain waters. And as uncertainty rises, the value of a well- *thought-out strategy drops. In fact, when pursuing entirely new business models, no amount of research can resolve the critical unknowns. All that strategy can do is give you a plausible starting point. From there, you must experiment, learn, and adapt.* (Govindarajan and Trimble, 2004)

The first step in past and current efforts to introduce conservation on agricultural landscapes is planning. The conservation plan has been the cornerstone and focus for most conservation programs in the United States. But as Govindarajan and Trimble (2004) point out in the quotation above, it is not the idea or the plan but the execution that produces success. In other words, a plan alone, even accompanied by innovative technology, is not complete without significant attention also being given to execution.

Manure management on the farm is a good example of the difference between planning outcomes and achieving those outcomes. Livestock producers in the United States must comply with the 1999 USDA-U.S. Environmental Protection Agency (EPA) United National Strategy for Animal Feeding Operations (AFOs). That strategy specified that all AFOs should develop and implement a technically sound, economically feasible, and site-specific nutrient management plan to minimize impacts on water quality and public health. An important issue for AFOs and nutrient managers is the level of available phosphorus in the soil because phosphorus is a major water quality pollutant. Soil phosphorus generally increases with historical applications of manure; therefore, manure application rates, locations, and timing are integral to the comprehensive nutrient management plan (CNMP). Using the NRCS 590 nutrient management practice standard and Wisconsin technical note Conservation Planning WI-1, a manure management plan was written for the period from September 1, 2000, to September 1, 2001, for an AFO managing about 200 dairy cows. Each field was assigned an application rate, in this case primarily based on soil-test phosphorus (Bray-1 phosphorus) and yield potential (Figure 1a). Cabot et al. (2006) tracked manure application patterns on this dairy farm to determine where manure was actually applied. This mapping of manure application locations (Figure 2) indicated that CNMP recommendations were not adhered to at many locations.

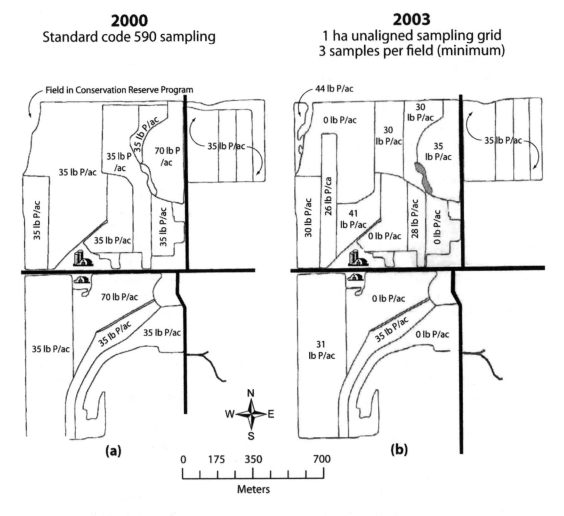

Figure 1. Comprehensive nutrient management plans for a dairy farm based on (a) standard guidelines and (b) 1-hectare-grid soil phosphorus tests. Source: Cabot, unpublished.

Another critical dimension is the scale at which a plan is developed and executed. Intensive soil sampling [1-hectare resolution (2.5 acres)] was done in spring 2003 to re-measure soil-test phosphorus at a finer spatial scale (Cabot and Nowak, 2005). A new phosphorus recommendation map, based on the intensive soil-test phosphorus data, was then written for the period from August 31, 2003, to September 1, 2004, showing that the phosphorus fertilizer recommended for most fields declined and in some cases reached zero (Figure 1b). Unfortunately, execution of this more intensive plan at the field scale will probably not achieve the desired environmental benefits.

The fact is that both the 2000-2001 and 2003-2004 nutrient management recommendations were based on the field scale, without addressing how the operator actually applied manure to the landscape. Neither did the recommendations address disproportionalities in the potential for phosphorus movement on the landscape (Nowak and Cabot, 2004; Nowak et al., 2006). CNMPs ought to be based on landscape vulnerabilities to phosphorus transport that are consistent with how livestock operators actually execute the distribution of manure on the landscape, according to principles outlined by Bennett et al. (1999), Carpenter (2005), and Sharpley et al. (2001).

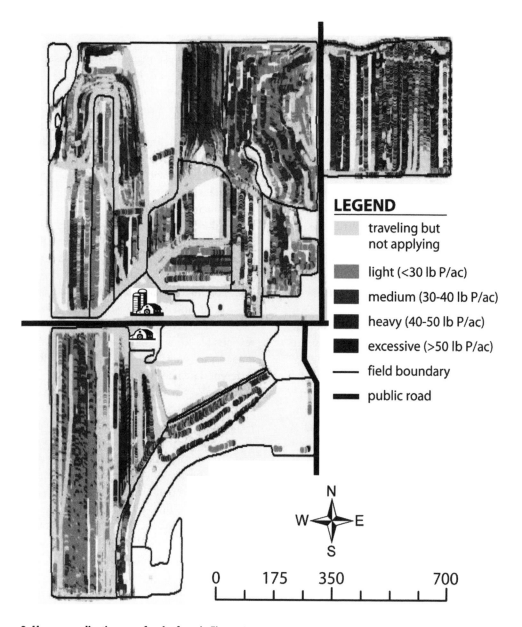

Figure 2. Manure application map for the farm in Figure 1.

Profiling users—users have different data needs

Three distinct groups generate and use data on managing agricultural landscapes: Scientists, program managers, and problem-solvers. Each group has different data needs. Those needs vary by type of data, level of detail, and spatial and temporal scale. Scientists are oriented to the scientific process and strive to quantify, explain, and manage variability in a fashion acceptable to peers through review and publication. Their metric is variability explained. Program managers require data and information to manage the people, fiscal resources, and political accountability associated with conservation programs. Their metrics are associated with the accountability measures often linked to those programs. Problem-solvers are not interested in explaining variation or managing programs, but rather in data needed to make appropriate management decisions that address or solve everyday problems. Their metrics are associated with an increase in value that results from solving or addressing a problem. The dilemma is that data needed to meet the expectations of these different groups are rarely specified in the landscape management strategy. Instead, they are mixed together, confounded, and often incomplete as a result for one or more of the groups.

Moreover, it is rarely acknowledged that the data needs of these different groups are often incompatible with each other. The data associated with a scientific analysis may have little utility to a problem-solver who wants to increase net income, and most scientists will see little value in knowing how much money was spent in a watershed, the number of plans developed, or the extent a practice was installed relative to the specifications in a field office technical guide. Part of the data dilemma is the untested assumption that scientific data translates into policy that translates into implementation guidelines that will solve problems. One could argue that the implicit direction in these data relationships could be reversed. That is, create a diverse set of arrangements where local resource professionals and landowners have the capacity to solve local problems. Then create programs around those processes, procedures, and practices that appear to be most effective and efficient for solving problems. Finally, use science to validate conservation solutions and determine where or when in the agricultural landscapes alternative solutions are needed.

Viable resource management across agricultural landscapes that enhances ecological goods and services while optimizing productivity and profits will require different types of data. Scientific data generated to explain and predict variation across different spatial and temporal scales remains a critical data need. In addition, it is suggested that additional emphasis be given to the data needs of the program manager and local problem-solver. An important dimension of our data dilemma has been the failure to recognize that data requirements associated with explaining variation in a problem is not the same as data requirements for managing a program created to address the problem, which, in turn, is not the same as data needed to solve the problem.

Summary and conclusions

If you can't measure something, you really don't know much about it. If you don't know much about it, you can't control it. If you can't control it, you are at the mercy of chance. (Bhatia, 2005)

Data and modeling requirements needed for a shift from managing fields to managing agricultural landscapes will be considerable. The data dilemma is that these multiple and diffuse data requirements are complex, yet receive little attention in the discussion and analysis of how to manage agricultural landscapes. We believe this data dilemma can be addressed by more creative use of technological innovation, focusing data processes on the critical portions of the landscape, structuring and organizing the data to encourage execution or action, and recognizing that those responsible for improving environmental performance across agricultural landscapes will have different data needs.

Technology will be a double-edged sword in addressing the data dilemma. On the one hand, technology will allow us to collect large amounts of data across multiple spatial and temporal scales. Technological innovation will also create the opportunity to collect data automatically on parameters that are currently labor-intensive, expensive, and time-consuming. But this same technological innovation will also generate massive amounts of raw data that will in many cases overwhelm current analytic capabilities. Moreover, it is not just the scientist who will face this situation of data-overload. Large amounts of high quality data, both biophysical and socioeconomic, will be generated on the farm, will be owned by the farmer or land manager, and could be used for production decisions and adaptive management as part of the process of enhancing value. This will require many farmers to seek out technical-service providers for internet data management and analysis, while others who cannot afford or do not have access to those services will fall into the "data divide." Program managers in this situation may retreat to the myopic position of focusing data analysis on efficiency measures (i.e., cost per unit of program action) while ignoring effectiveness measures (i.e., extent program goals are achieved). There is also the danger of local resource management professionals becoming office-bound data managers rather than practicing conservationists.

The capacity exists to improve environmental performance significantly on agricultural landscapes. A major obstacle to achieving this potential is the data dilemma. More, better, or technologically enhanced data by itself is not the answer. Data must be transformed into information, and information must be managed to create or enhance value. Using this foundation, a strategic analysis of what data is gathered and how that data is gathered and analyzed is sorely needed. The process must recognize the different data needs of different data users. This is the challenge confronting programs like CEAP.

References

Bennett, E.M., T. Reed-Andersen, J.N. Houser, J.R. Gabriel, and S.R. Carpenter. 1999. A phosphorus budget for the Lake Mendota Watershed. *Ecosystems* 2(1): 69-75.

Bhatia, A. 2005. Data rich, information poor? Focus on right metrics. The Data Administration Newsletter (TDAN.com). http://www.tdan.com/i034hy03.htm.

Cabot, P.E., and P. Nowak. 2005. Planned versus actual outcomes as a result of animal feeding operation decisions for managing phosphorus. *Journal of Environmental Quality* 34: 761–773.

Cabot, P.E., F.J. Pierce, P. Nowak, and K.G. Karthikeyan. 2006. Monitoring and predicting manure application rates using precision conservation technology. *Journal of Soil and Water Conservation* 61(5): 282-292.

Carpenter, S.R. 2005. Eutrophication of aquatic ecosystems: Biostability and soil phosphorus. *Proceedings, National Academy of Sciences* 102(29): 10,002-10,005.

Carpenter, S.R., R. DeFries, T. Dietz, H.A. Mooney, S. Polasky, W. V. Reid, and R.J. Scholes. 2006. Millennium ecosystem assessment: Research needs. *Science* 314(5797): 257-258

Edwards, W., and L. Owens. 1991. Large storm effects on total soil erosion. *Journal of Soil and Water Conservation* 46: 1, 75-78.

Fenn, M.E., J.S. Baron, E.B. Allen, H.M. Rueth, K.R. Nydick, L.Geiser, W.D. Bowman, J.O. Sickman, T. Meixner, D.W. Johnson, and P. Neitlich. 2003. Ecological effects of nitrogen deposition in the western United States. *Bioscience* 53(4): 404-420.

Galloway, J., J. Aber, J. Erisman, S. Seitzinger, R. Howarth, E. Cowling, and B. Cosby. 2003. The nitrogen cascade. *Bioscience* 53(4): 341-356.

Goodwin, S. 1996. Data rich, information poor (DRIP) syndrome: Is there a treatment? *Radiological Management* 18(3): 45-9.

Govindarajan, V., and C. Trimble. 2004. Strategy, execution, and innovation. Fast Company. August. Accessed January 15, 2007. http://www.fastcompany.com/resources/leadership/vgct/081604.html.

Grace, J., P.R. van Gardingen, and J. Luan. 1997. Tackling large-scale problems by scaling-up. p. 7-16. In P.R. van Gardingen, G.M. Foody, and P.J. Curran, editors, Scaling-up from Cell to Landscape. Cambridge University Press, New York, New York.

H. John Heinz III Center for Science, Economics, and the Environment. 2002. The state of the nation's ecosystems: Measuring the lands, waters, and living resources of the United States. Cambridge University Press, Cambridge, United Kingdom.

Johannsen, C.J., C. Clark, T. David, A. Falconer, M. Goodchild, D. John, F. Pierce, H. Schreuder, N. Tosta,R. Unnasch, M. Vanacht, and J. Zinn. 1995. Data rich and information poor. Blue Ribbon Panel on Natural Resources and Performance Measurement. Natural Resources Conservation Service, U.S. Department of Agriculture, Washington, D.C. 36 pp.

Kurzweil, R. 2001. The law of accelerating returns. Accessed December 18, 2006. http://www.kurzweilai.net/articles/art0134.html?printable=1

Millennium Ecosystem Assessment. 2005. Living beyond our means: Natural assets and human well-being. Statement from the board. Available on-line at http://www.millenniumassessment.org. World Resources Institute, Washington, DC.

Moore, G.E. 1965. Cramming more components onto integrated circuits. *Electronics Magazine* 38(8):114-117.

Nowak, P., and P.E. Cabot. 2004. The human dimensions of resource management. *Journal of Soil and Water Conservation* 59(6): 128-135.

Nowak, P., S. Bowen, and P.E. Cabot. 2006. Disproportionality as a framework for linking social and biophysical systems. *Society and Natural Resources* 19:153-173.

Philips, S.W. 2005. U.S. Geological Survey Chesapeake Bay science plan, 2006-2011. Open-File Report 2005-1440. U.S. Geological Survey, U.S. Department of the Interior, Reston, Virginia.

Pierce, F.J., J. L. Chavez, T.V. Elliott, G. Matthews, R.G. Evans, and Y. Kim. 2006. A remote-real-time continuous move irrigation control and monitoring system. Paper No. 062162. American Society of Agricultural and Biosystems Engineering St. Joseph, Michigan.

Pierce, F.J., and T.V. Elliott. 2007. Regional and on-farm wireless sensor networks for agricultural systems. *Computers and Electronics in Agriculture* (accepted).

Pierce, F.J., and P. Nowak. 1999. Aspects of precision agriculture. *Advances in Agronomy* 67:1-85.

Raun, W., and G. Johnson. 1999. Improving nitrogen use efficiency for cereal production. *Agronomy Journal* 91: 357-363.

Raun, W.R., J.B. Solie, G.V. Johnson, M.L. Stone, R.W. Mullen, K.W. Freeman, W.E. Thomason, and E.V. Lukina. 2002. Improving nitrogen use efficiency in cereal grain production with optical sensing and variable rate application. *Agronomy Journal* 94: 815-820.

Raun, W.R., J.B. Solie, M.L. Stone, K.L. Martin, K.W. Freeman, R.W. Mullen, H. Zhang, J.S. Schepers, and

G.V. Johnson. 2005. Optical sensor based lgorithm for crop nitrogen fertilization. *Communication in Soil Science and Plant Analysis* 36: 2,759-2,781.

Sharpley, A., R. McDowell, J. Weld, and P. Kleinman. 2001. Assessing site vulnerability to phosphorus loss in an agricultural watershed. *Journal of Environmental Quality* 30: 2,026-2,036.

Sharpley, A.N., P. Kleinman, and R. McDowell. 2001. Innovative management of agricultural phosphorus to protect soil and water resources. *Communications in Soil Science and Plant Analysis* 32(7-8): 1,071–1,100.

Shipitalo, M., and L. Owens. 2006. Tillage system, application rates, and extreme event effects on herbicide losses in surface runoff. *Journal of Environmental Quality* 35: 2,186-2,194.

Soil and Water Conservation Society. 2006. Final report from the Blue Ribbon Panel conducting an external review of the U.S. Department of Agriculture's Conservation Effects Assessment Project. Ankeny, Iowa. 26 pp.

Ward, R.C., J.C. Loftis, and G.B. McBride. 1986. The "data-rich but information-poor" syndrome in water quality monitoring. *Environmental Management* 10(3): 291-297.

Part **3**

The science of targeting
within landscapes
and watersheds to
improve conservation
effectiveness

The science of targeting within landscapes and watersheds to improve conservation effectiveness

Todd Walter

Mike Dosskey

Madhu Khanna

Jim Miller

Mark Tomer

John Wiens

Certain portions of the landscape are especially sensitive to human activities or strategically located to mitigate the environmentally detrimental impacts of human activities. Targeted land management is thus defined here as the focusing of preservation, conservation, or other practices on those specific portions of the landscape (or at particular times) where (and when) they will have the greatest benefits at the lowest economic costs (assuming that conservation practices actually reduce net economic return).

Motivation and need for targeted management are increasingly acute as society's pressures on and valuing of natural resources simultaneously increase. Ideally, targeted land management allows conservation funds to improve environmental quality most effectively and balance a complex and shifting mosaic of land uses that vie for space and differ in the economic returns they provide. In the case of land conservation programs, this funding is often needed to compensate landowners for adopting environmentally friendly practices that involve foregoing some or all of the economic returns to land.

The "targeting" concept evolves around improved scientific understanding of how ecological, hydrological, and geomorphological processes and economic costs of land management are distributed and interact across the landscape and over time. These advances in knowledge of ecological and earth-system connectivity have paralleled improvements in geographic information systems (GIS), remote sensing, and other methods of gathering, compiling, and analyzing geospatial data. Because of the increasingly complex suite of management objectives, effective implementation requires scientifically based planning tools, guidelines, or strategies to match practices and locations with optimal effectiveness.

To help focus this discussion, we primarily consider spatial rather than temporal aspects of targeting, for which the reader is referred to other sources (e.g., Edwards et al., 1992; Randall and Mulla, 2001; Dinnes et al., 2002). Here then, targeting is usually the process of identifying priority locations for implementing conservation or habitat improvement practices. Ideally, targeted management incorporates the combined or integrated effect of many individually targeted practices (e.g., Vieth et al., 2003, 2004). Similarly, targeted landscape management often balances multiple objectives, usually protecting water quality, soil health, terrestrial or aquatic habitat, and/or public safety while simultaneously recognizing economic and other management issues that may be ownership-specific. Planners and decision-makers may choose among a variety of balancing criteria, such as maximizing environmental benefits achieved, minimizing economic costs of achieving environmental benefits, and maximizing environmental benefits subject to some cost or land-area constraint, among others. The choice of objectives is important to implementing a targeted management plan because the correlation between the distribution of environmental benefits and economic costs of land parcels within a region plays an important role in deciding which land parcels get targeted (Babcock et al., 1996, 1997). For purposes of this discussion, targeting seeks to optimize the beneficial effects of conservation practices on natural resources with a given conservation budget and thereby maximize the returns on economic investments in conservation.

Targeting as a process is not restricted to any particular scale of space or time. The scale at which we target land management depends upon the interplay among a number of factors, including specific management objectives, relevant ecological or earth-system processes, the precision at which management decisions can be implemented, the resolution of available data, and potential sociopolitical constraints. One can find examples of targeted management across a wide range of scales, from very large areas (e.g., preservation of terrestrial habitats) and whole watersheds (e.g., U.S. Environmental Protection Agency hi-priority watersheds) to individually owned parcels (e.g., nutrient management plans for farms) or smaller areas (e.g., riparian areas or stormwater infiltration basins).

The case for targeting

There are three compelling reasons for promoting targeted land management to meet resource conservation or preservation goals. First, agriculture is a "landscape" enterprise, of which the fields in agricultural production are only a part. Landscapes are, by definition, heterogeneous. If all places in a landscape were the same or if we could consider all places (agricultural fields, streams, watersheds) independent of their surroundings, there would be no need for targeting. But the context of agricultural fields (even seemingly homogenous ones) in a spatially varied landscape creates the need for and the opportunity to use targeting to improve the effectiveness of land use and conservation management.

Second, there is irrefutable scientific evidence that some locations in the landscape have a high pollutant-generating potential (sensitive sites), can function effectively to intercept and treat pollutants, and/or have features that comprise critical habitat for wildlife. By understanding how such processes or properties vary spatially, efforts to mitigate their detrimental effects or enhance their beneficial consequences can be applied more precisely.

Third, the economic costs of conservation practices also differ across locations. Accordingly, targeting allows cost-effective conservation efforts to focus funds for resource conservation in locations where consequences, such as pollution-reduction potential, are high and costs are low (Johansson and Randall, 2003).

It is not possible to apply conservation treatments across the entire landscape. Ecosystems cannot all be entirely reconstructed, and we must accept that human activities impact soil, water, and habitat resources. Therefore, methods are needed to identify and protect areas that are sensitive (most prone to environmental damage), able to effectively mitigate pollution, and/or provide critical habitat. The decision to adopt a conservation practice to protect the environment is most effective if it is a voluntary decision by a landowner, rather than one mandated by government. Ideally, such decisions would be based on increased social consciousness toward environmental protection through education and technical assistance that encourages landowners to adopt conservation practices. Moral suasion and efforts to raise this kind of consciousness, however, have an insignificant effect in promoting adoption of conservation practices that are unprofitable unless the landowner's immediate environment or health is at risk. In contrast, economic incentives through cost-sharing do have a significant impact on conservation behavior (Batie and Ervin, 1999). Designing market-based incentives for landowners to adopt conservation practices, therefore, has potential for improving participation in environmental protection.

The key questions in designing and implementing conservation policies that employ targeting, then, are (1) how do we develop methods that identify specific targeting criteria for each resource, especially in the context of multiple-criteria objectives and (2) how do we translate these criteria into conservation planning tools/instrumentation that can be developed at regional or watershed scales and applied at the farm-field (ownership) scale in ways that are cost-effective across multiple agricultural producers?

Targeted management: An evolving concept

Clearly, there are historical precedents for targeting conservation practices. Indeed, the National Park System is a form of targeted environmental protection. Also, agricultural producers intuitively understand targeting and have, for decades, routinely timed fertilizer applications to maximize crop uptake (Dinnes et al., 2002) and, more recently, utilized precision agricultural technologies to focus fertilizer applications on parts of the landscape where crop nutrient utilization will be maximized (Rejesus and Hornbaker, 1999; Kaspar et al., 2003).

Since the 1930s, U.S. Department of Agriculture (USDA) programs and activities have encouraged farmers to establish conservation practices on environmentally sensitive land. While primarily based on localized knowledge up through the 1970s (Heimlich, 2003), the 1985 farm bill and USDA's Conservation Reserve Program (CRP) more clearly defined areas that should be targeted to protect wetlands and reduce soil erosion. Specifically, USDA's universal soil loss equation, in conjunction with site-specific soil survey information, was used to define quantitatively highly erodible land (Benbrook, 1988) and a federal interagency definition of "wetland" was developed with hydric soils as a key indicator (Heimlich, 1994). More recently there has been an important, implicit shift from on-site impacts to down-system influences, for example, water quality impairment attributed to up-slope land management.

Concerns continue to broaden beyond soil erosion (highly erodible land) and water storage (wetlands) to include water quality, air quality, and wildlife. The number of governmental programs has expanded, and eligibility criteria for conservation treatments have become more specific. For example, the environmental benefits index developed for evaluating and ranking land parcels offered for CRP enrollment includes a detailed evaluation of potential for wildlife habitat and water quality improvement, in addition to protection of highly erodible soils (U.S. Department of Agriculture-Farm Service Agency, 1999) and the soil rental rate of each parcel. This index is based on landscape information, including highly erodible soils, proximity to water and wetlands, and location of state-designated critical areas for water quality (e.g., Clean Water Act Section 303d listed waters) or wildlife (e.g., habitat areas for species listed under the Endangered Species Act). These indices are typically used to rank conservation proposals from pools of voluntarily submitted applications to prioritize incentive payments. Despite such use of these indices, increasingly complex eligibility determinations still must be implemented on the basis of professional judgment and interpretations of general soil and landscape position rather than on detailed, robust scientific understanding of landscape functions, processes, and quantitative relationships.

Over the past decade, local planning efforts have become important driving forces behind targeted landscape management, especially with respect to watershed management for water quality and aquatic habitat protection (e.g., Walter and Walter, 1999; House, 2000; hundreds of plans are not widely publicized). Local officials are generally eager for sound scientific bases to prioritize water quality improvement projects. One high-profile example is the New York City (NYC) Agricultural Watershed Protection Project (a.k.a., Watershed Agriculture Program/WAP; Walter and Walter, 1999), which is directed by a Watershed Agriculture Council (WAC) largely made up of local producers and landowners who partner with regional research universities to develop the underpinning science for targeted management practices.

Some states have developed formal legislation to promote these types of grassroots efforts, such as Washington State's 1998 Watershed Planning Act, which states "…the local development of watershed plans for managing water resources… is vital to both state and local interests." Indeed, this type of localization of management allows scientists to develop targeted management strategies that match local or regional conditions and controlling processes in ways that can be more difficult under broader, usually nationally mandated efforts. Explicitly included in the NYC watershed program was a requirement to "test, demonstrate, and evaluate the scientific…criteria developed for whole farm planning (New York State Water Resources Institute, 1992)," most of the components of which were various targeted best management practices (BMPs) at farm and field scales for controlling nonpoint-source pollution in NYC water-supply reservoirs (e.g., Walter et al., 2000, 2001).

Although there has been no formal analysis, one common, implicit theme among successful watershed projects is a close interaction between landowners, planners, and/or researchers (e.g., Meals, 2001; Wang et al., 2002; Dietz et al., 2004; Bishop et al., 2005; Lovegreen et al., 2006). Those types of locally driven efforts also facilitate accommodating the differences in priorities, concerns, and objectives among individual landowners, local planners, and broader state and national governments.

It appears that current trends in targeted land management are a natural part of the progression of environmental conservation history. Hopefully, the shift over the past decade or so toward locally initiated conservation and protection efforts will continue to redefine the role of top-down policies. For example, the Conservation Reserve Enhancement Program (CREP), a federal-state partnership program initiated in 1996, has helped regionalize cost-effective targeting of high-priority, environmentally sensitive land based on quantitatively defined environmental goals (U.S. Department of Agriculture-Farm Service Agency, 2003), although it could be improved to better engage grassroots activities. As discussed earlier, increased local control also potentially facilitates closer ties between planners or researchers and landowners, ensuring better application of scientific concepts to targeting efforts.

Although targeting allows planners to realize the continuum of objectives and natural processes distributed across the globe, we focus here on water quality and terrestrial habitat protection for wildlife, two prominent objectives that illustrate the breadth of science, technology, and future challenges and opportunities of targeted land management.

Water quality protection and enhancement

The state of the scientific support for targeting

An important, recent shift in targeting of conservation practices is an improved understanding of how individual parts of the landscape are interconnected. Early water quality protection efforts involved little more than a repackaging of soil conservation practices (Walter et al.,

1979; Walter et al., 2000). Although our efforts to gain knowledge of critical processes and their scales will surely continue to challenge us, scientific advances during the past 30 years have provided ample justification for targeted water quality protection practices beyond those currently employed as BMPs. In particular, there have been substantial advances in understanding how hydrological processes are distributed across the landscape (e.g., Hewlett and Hibbert, 1967; Dunne and Black, 1970a,b; O'Loughlin, 1981; Moore et al., 1991; Lyon et al., 2006a,c) and how those processes are influenced by soil variability (e.g., Thompson et al., 1997; Gessler et al., 2000; Western et al., 2004) and correlate with crop productivity (e.g., Kaspar et al., 2003; Kravchenko and Bullock, 2000).

Targeting of conservation practices is increasingly based on those advances in distributed hydrology and sound scientific understandings of relevant ecological and/or transport processes. Recent scientific, paired-watershed evaluations (Loftis et al., 2001), although somewhat nonspecific, suggest that a targeted approach can effectively improve water quality and aquatic health, especially practices that protect riparian areas (e.g., Meals, 2001; Wang et al., 2002; Dietz et al., 2004; Bishop et al., 2005). Studies to evaluate rigorously the effectiveness of targeted versus other nontargeted conservation implementations have not been conducted, however, nor do we have a good understanding of the system-wide impacts of specific implementations and/or maintenance of particular practices and combinations of practices. In lieu of good field evaluations, we have relied (perhaps too much) on model results to demonstrate the effectiveness of targeted management (e.g., Walter et al., 2001; Veith et al., 2004; Mankin et al., 2005). More reliable evaluations will require innovatively designed, long-term, and/or paired-watershed studies implemented at the appropriate scale. Furthermore, linkages between conservation practices and water quality response in large watersheds must bridge the scales of implementation (field) and measurement (watershed) or risk being experimentally flawed (see Gardner et al., 2001).

This bridging of scales is challenging, but the risk of ignoring this challenge is clearly illustrated, for example, by the recognition that soil conservation practices have not always resulted in anticipated reductions in stream sediment loads (e.g., Trimble, 1999a,b; Trimble and Crosson, 2000a,b). This is because of the widely employed, simplistic assumption that the amount of sediment generated by a land parcel and the amount reaching a water body depend only upon a fixed proportion or the distance of the parcel from the water body (e.g., Khanna and Farnsworth, 2005). So, although the scientific rationale for many (often targeted) soil conservation practices is sound and those practices do arrest excessive soil loss, the intended benefits for rivers have not been recognized. That is because we did not fully understand or consider the integration of practices, processes, and expected environmental benefits and thus failed to select appropriate experimental designs and monitoring scales. Experiments and monitoring to demonstrate water quality benefits from targeted practices need be designed to consider the landscape position of land parcels and consider how processes are distributed along flow paths; in this example, upland erosion, redeposited, and channel-bed erosion (Nagle and Ritchie, 1999, 2004; Nagle et al., 2006). Interestingly, Khanna et al. (2003) showed that retiring particular land parcels without considering the management decisions of surrounding parcels considerably underestimates the resulting sediment loads to streams, at much higher costs than targeted practices that consider land parcels in their hydrologic context. This soil erosion, stream sediment example also demonstrates the risk of relying too much on models to make predictions, that is, our overly simplistic assumptions of how parcel- or plot-scale erosion is linked to sediment delivery to streams were built into the models, which gave poor estimates of how targeted soil-conservation practices would improve stream water quality (e.g., Trimble, 1999b; Trimble and Crosson, 2000b).

In summary, the scientific rationale for many targeted practices is good. Indeed, the past decade has seen a notable shift toward targeted management strategies using process-based scientific understanding of these practices. Subsequent studies, although too few, suggest that the implementation of those practices is having positive impacts on environmental quality (e.g., Meals, 2001; Wang et al., 2002; Dietz et al., 2004; Bishop et al., 2005; Lovegreen et al., 2006). But we must continue to improve our understanding of how

processes are integrated over whole systems at different scales, which will ultimately allow us to better evaluate targeted management successes and failures in ways that illuminate knowledge gaps and meaningfully direct future progress. It is important to remember, however, that the scientific bases for targeting conservation practices need to interface ultimately with planners and land managers in ways that combine local/regional knowledge and objectives to develop better and more economical environmental protection practices. In essence, an improved scientific basis for targeting of practices is allowing planners to move increasingly away from "one size (scale) fits all" management approaches. We comment further on the issue of scaling targeting in the concluding section of this paper.

The state of targeting technology

Tools for targeting biogeo processes. Technological advances in monitoring and managing increasingly complex, spatially distributed information and in precisely locating landscape positions have enabled revolutionary advances in targeted land management capabilities. While we do not want to diminish the contributions these technological advances have made to better targeted water quality protection and enhancement, we think it is arguably more important to recognize that many targeting approaches to protect water quality are based on decades-old data (e.g., soil surveys), methods (e.g., soil phosphorus tests), and models (e.g., curve number runoff model, U.S. Department of Agriculture-Soil Conservation Service, 1972), which have been recycled through user-friendly GIS in ways that were neither anticipated nor intended by their originators. These "old" tools and technologies generally worked well for their intended purposes, but we cannot expect them to be uniformly applicable for effective targeting of water quality management across all regions or large river basins. Scientifically, we should be able to state our assumptions about the utility of each data resource used for targeting, and we must understand the likelihood of those assumptions being invalid and implications of employing them erroneously. In short, if targeting is to become part of policy one day, then at some point targeting criteria may need to be defended in court.

To illustrate this issue, one pervasive example is that among modelers and environmental protection practitioners there may be an implicit assumption that electronic geospatial data all have similar levels of precision or scientific reliability. While some data, like the newest digital elevation data (e.g., LIDAR), are very detailed and their errors well documented, others, especially soils data, are generally digitized forms of hand-drawn maps. Of course, soil scientists mapping North America's agricultural soils, largely during the mid-20th century, understood that soils are expressed as a landscape continuum, but mapping capabilities at the time required them to delineate general groups of soil characteristics by polygons on a map. While this traditionally discrete format for representing soil properties has been historically convenient for soil conservation targeting (e.g., identifying highly erodible land) or wetland identification (as indicated by hydric soils), these same data are incorporated into increasingly sophisticated analyses and mechanistic models without questions about their accuracy, precision, or variability, which can be significant from region to region. Despite this, some process-based hydrologic models perform well using these soils data (e.g., Frankenberger et al., 1999; Mehta et al., 2004; Gerard-Marchant et al., 2006; Schneiderman et al., 2006; Easton et al., 2006), suggesting that the current level of precision of soil survey data is adequate for targeting applications in at least some regions.

Similarly, there is an intriguing misconception that currently used water quality models, such as the soil and water assessment tool (SWAT) (e.g., Arnold et al., 1993), agricultural nonpoint-source pollution model (AGNPS) (Young et al., 1989), and the generalized watershed loading function (GWLF) (Haith and Shoemaker, 1987), provide highly targeted management insights, presumably because they use geospatially referenced data and the resulting model output can be presented as a digital map. Unfortunately, these models typically rely on the USDA Soil Conservation Service [now the Natural Resources Conservation Service (NRCS)] curve number equation (e.g., U.S. Department of Agriculture-Soil Conservation Service, 1972) to predict runoff. Although tables are available that link runoff potential (i.e., curve numbers) to specific land uses, there is little reliable scientific basis for these indices of runoff

potential, especially for small storms (Walter and Shaw, 2005). Indeed, the model's creator, Victor Mockus, justified his model largely "on grounds that it produces rainfall–runoff curves of a type found on natural watersheds" (Rallison, 1980). So while the curve number method may provide a reasonable estimate of watershed-scale runoff, and, indeed, has been a useful engineering design tool for many decades, in most cases it cannot predict runoff from a specific location within a watershed. As currently used, then, water quality models may be reasonably effective at targeting priority watersheds, but are unlikely to target specific locations within a watershed for implementing BMPs or other water quality protection strategies (e.g., Garen and Moore, 2005). On a positive note, there have been substantial efforts to reconceptualize the traditional curve number equation in ways that are scientifically defensible for particular situations (e.g., Steenhuis et al., 1995; Gburek et al., 2002; Lyon et al., 2004; Scneiderman et al., 2006).

There are also several other terrain analyses that have potential applicability for improving the physical basis of water quality models with respect to specific runoff processes at hillslope scales (Moore et al., 1991; Bren, 1998; Gburek et al., 2002; Tomer et al., 2003; Agnew et al., 2006; Lyon et al., 2006b). Although more sophisticated process-based models may eventually provide a more scientifically robust hydrologic basis for water quality models, for the most part they are currently too cumbersome for practical use. In the foreseeable future it appears that institutional momentum will perpetuate the use of current water quality models and, hopefully, practitioners will adopt approaches for improving the scientific basis and decision-support capacities of these models.

We do not have space to discuss similar, potentially problematic reinventions of traditional data and approaches (for example, soil phosphorus tests originally developed to assess cropland productivity are now commonly used to assess phosphorus loading to streams). As with improvement in digital elevation data (e.g., LIDAR), it is likely that the accuracy and precision of other geospatial data will continue to improve as new environmental monitoring technologies mature. In the short term, it is important to recognize explicitly the gap between scientific advances and current technologies and tools for water quality "targeting." While the greater user-friendliness of these tools has improved their accessibility, it also risks facilitating their misuse. Thus, effective use of these tools could be improved by developing guidelines that explicitly address their technological limitations. That probably requires the professional judgment of specialists in combination with local knowledge.

Economic instruments to support targeting. Identification of land parcels that should be targeted for improved management to achieve environmental benefits at least cost can be achieved, at least in principle, by integrating an economic model with a water quality model, together with detailed GIS information on the characteristics of land parcels. Such integrated models typically assume that the economic costs of adopting conservation practices by a landowner are known and observable to a policy planner. They generally ignore the possibility of asymmetric information because the true cost of adopting a conservation practice is private information known only to the landowner. They also typically ignore the possibility of moral hazard: A landowner may adopt a conservation practice, but not make a full commitment to implementing it. Nevertheless, economic models that incorporate spatial heterogeneity in costs and physical processes and the implicit interdependencies of associated sediment-abatement benefits can improve policy planners' abilities to target conservation practices to reduce offsite pollutant loadings (Khanna et al., 2003).

Targeting, as referred to here, is defined normatively and from the perspective of a policy maker: It identifies land parcels on which conservation practices should be adopted to achieve environmental goals most cost-effectively. Because conservation efforts by landowners represent voluntary decisions in response to market-based incentives, policymakers need to design a "green payment" policy, such as subsidies for adoption of a conservation practice on cropland or a rental payment for retiring a land parcel from crop production, to provide incentives for landowners to adopt costly conservation practices on the targeted land parcels. Integrated, spatially explicit economic models can be used to design these per-acre green payments. In the presence of spatial heterogeneity in costs and environmental benefits,

such payments may need to be parcel-specific in order to achieve conservation goals. For example, to achieve sediment-abatement goals through land retirement most cost-effectively, economic models show that the per-acre rental payments to landowners should vary with the location of the parcel relative to water bodies, the quality of the soil, the slope of the land parcel, and the soil erodibility index. To implement these site-specific rental payments, policymakers or planners need to know the relationships and parameters embedded in the integrated model. This is information- and skill-intensive. In practice, conservation programs tend instead to adopt second-best approaches to target land use change. For example, soil rental payments offered for enrollment in the CRP are soil-specific, but do not vary with the environmental benefits provided by that parcel. As another example, the Illinois CREP targets environmentally sensitive land by limiting eligibility for enrollment in the program to land parcels within a narrowly defined area in the Illinois River Basin. But it does not specify any mechanism to select 132,000 acres from the 7 million acres of heterogeneous land parcels in this area, which may ultimately raise program costs (Yang et al., 2004). Similar issues are prevalent with this program across the United States.

Challenges and future directions

What is needed to advance the targeting of water quality management and policies? Foremost, national targeting policies are needed that promote linkages between local, grassroots efforts and regional scientific researchers in ways that bridge science-application gaps. Interestingly, this need is echoed in one of the earliest discussions on targeting conservation practices by Maas et al. (1985), who emphasized that setting targeting criteria should take place within the context of watershed-specific planning. In other words, scientific knowledge must be adapted locally to set targeting criteria that are useful for site-specific decision-making; thus, human judgment is an important part of this process. For example, setback distances from water bodies must combine scientifically defensible criteria (e.g., Tomer et al., 2003, Walter et al., 2005) with local knowledge that is not necessarily applicable across broad regions. Indeed, this type of policy speaks directly to the problems associated with tensions between historical approaches and new technologies. Presumably, scientists are aware of these potential problems and can help local planners meaningfully interpret data and model results that may have "hidden" limitations, which might only be revealed through feedback from landowners or field reviews. Scientists are also uniquely qualified to recognize controlling processes and, thus, appropriate analytical approaches. For example, in some regions identifying hydrologically active areas for runoff generation (specifically, variable-source areas) is important to targeting water quality management practices. There are several scientifically defensible methods for identifying these areas based on topography (e.g., O'Loughlin, 1986; Vertessy et al., 1999; Mehta et al., 2004; Agnew et al., 2006), although these must be interpreted in the context of the planning objectives and the accuracy of available topographic data (Kuo et al., 1999; Tomer et al., 2003; Tomer and James, 2004), both of which can vary substantially from region to region.

A potential, though untested, dimension to this policy is that targeted management may be inherently more successful when scientists, planners, and landowners work closely together to establish linkages of trust. For example, Cornell researchers and the New York State Department of Environmental Protection worked for more than 10 years with an agricultural producer to target BMPs effectively in a small watershed in Delaware County and demonstrated substantial (about 40 percent) reductions in phosphorus and sediment loading with targeted land management (Bishop et al., 2005). Scientists and planners at the Bradford County, Pennsylvania Conservation District demonstrated similar pollutant reductions in a small watershed after more than 10 years of close cooperation with producers to develop appropriate, targeted practices (Lovegreen et al., 2006).

Of course, extensive monitoring allowed researchers to evaluate the success of those projects, but such monitoring is rare. Linking researchers, planners, and landowners encourages the implementation of monitoring systems that can provide scientifically valid assessments of the impacts of combinations of targeted practices on local- and wide-scale environmental impacts. Only in this way can experience with targeted management provide important feedback for improving future targeted management.

Ideally, such efforts will implement distributed monitoring networks, in addition to measures at watershed outlets, to characterize how hydrologic and biogeochemical processes are distributed and interact across the landscape and in streams. Those processes interact to create "biogeochemical hotspots," parts of the landscape where pollutant or nutrient mobility or retention may be especially acute (McClain et al., 2003). Substantial background research supports the conclusion that biogeochemical processes are distributed or punctuated across the landscape (e.g., McClain et al., 2003; Welsh et al., 2005) and it is likely that some current targeting strategies could be more effective if they took these into account. One notable example is recent work showing that reclaiming incised urban streams reconnects the stream channel and riparian area in ways that increase denitrification (e.g., Groffman et al., 2002; Groffman and Crawford 2003). Improved understanding of how seasonal hydrologic cycles interact with microbial ecosystems reveals why wetlands and buffers act as nutrient (especially phosphorus) sources for some parts of the year and sinks during others (e.g., Dillaha et al., 1988; Carlyle and Hill 2001).

One important scientific challenge is determining the mechanics and transport roles of shallow, rapid, near-surface flows, such as those associated with lateral, preferential flowpaths (e.g., McDonnell, 2005). Subsurface preferential flow paths (Beven and German, 1982; Flury et al., 1994; Kung, 1990; Sidle et al., 2001) may be important within hillslopes (e.g., Wilson et al., 1990; Noguchi et al., 2001) and at watershed scales (e.g., Gatcher et al., 1998; Angier et al., 2005). There is good evidence that even compounds traditionally perceived as strongly adsorbed to soil, such as phosphorus, can move rapidly and deeply into the soil profile and then move laterally through natural preferential flowpaths (e.g., Gatcher et al., 1998) or artificial drainage systems (e.g., Geohring et al., 2001). Often, local landowners are the only sources of information about how and where artificial drainages are distributed through a landscape, reinforcing the need for scientists to work with local landowners to target land management practices.

Finally, creative research linking landscapes to streams through "tracer technologies" will enhance water quality targeting. These include genetic fingerprinting (e.g., Dombek et al., 2000;

Carson et al., 2001; Hartel et al., 2002, which may require genetic libraries on a watershed-specific basis; see Wiggins et al., 2003), natural isotopes (e.g., Christophersen et al., 1990; Genereux and Hooper, 1998; Kendall, 1998; McGlynn and McDonnell, 2003; Soulsby et al., 2003; Uhlenbrook and Hoeg, 2003), rare-earth elements (e.g., Lui et al., 2004; Kimoto et al., 2006), and anthropogenic sources of chlorofluorocarbons, sulfur-hexafluoride (e.g., Busenberg and Plummer, 2000; Browne and Gulden, 2005), and radionuclides (e.g., Wallbrink et al., 1999; Russell et al., 2001; Collins and Walling, 2002; Nagle et al., 2006). Innovative bio-nano-technological tracers (e.g., Mahler et al., 1998) can be used to identify pollutant sources and estimate the impacts of land use change on future trends in water quality. Through innovative combinations of these methods, researchers may be able to challenge the "nonpoint" concept, that is, the idea that we are incapable of identifying seemingly diffuse sources spread across the landscape. "Nonpoint-source pollution" may be more a label for our ignorance than a statement of reality. Ultimately, all pollutants have their sources in time and space.

Terrestrial habitat preservation for wildlife

The state of the scientific support for targeting

The previous section dealt with targeting to enhance water quality protection. The focus generally was on the diminishment of water quality by pollution, sedimentation, and the like. Stream systems and the factors affecting them are clearly multiscalar, and what happens upstream can have impacts at considerable distances from the sources (witness hypoxia in the Gulf of Mexico). Most water quality targeting, however, has been applied at local scales, often within individual landholdings or small watersheds. This is largely because governmental incentive programs for soil conservation and water quality improvement have been directed toward individual landowners. This is another reason why the engagement of landowners in targeting conservation efforts is so important.

For wildlife, the participation of local stakeholders is no less important. Because the objectives of wildlife conservation deal with species that have large ranges or migrate over vast distances, or with ecological systems whose dynamics are determined by regional as well as local factors, targeting efforts must be more explicitly multiscalar. But even though the scale of application of targeting may differ for different objectives, the importance of habitat in wildlife conservation carries across all scales.

Of course, the realization that species and ecological systems need habitat to survive and function was the foundation of natural history long before conservation and natural resource management emerged as recognizable disciplines, with their own professional societies, journals, and degree programs. Now, loss of habitat and fragmentation of what remains are widely considered to be primary threats to the persistence of populations and species (Forman and Godron, 1986; Dramstad et al., 1996; Wilcove et al., 1998). It would seem easy, then, to target "habitat" as the objective for protecting such populations and species. Indeed, the creation of nature reserves and protected areas is usually framed in terms of one habitat or another. Quite apart from debates about what "habitat" is or is not (e.g., Morrison and Hall, 2002), however, the recent emergence of landscape ecology, in tandem with the increasingly sophisticated use of spatially referenced GIS, has shown that this perspective by itself is inadequate and incomplete.

Similar to the earlier discussion regarding land parcels with respect to soil erosion, fragments of "habitat" are not independent areas floating in an inhospitable background matrix, but parts of a landscape mosaic of varied elements, with differing degrees of connectedness. Moreover, landscapes are not fixed and static in time, but rather dynamic, shifting, and changing as a result of natural disturbances and land use change (e.g., Turner et al., 1995; Dale and Haeuber, 2001; Theobald, 2005). And because the responses of different organisms to the landscape differ at different scales, as do different ecological processes, a "one size (scale) fits all" approach to protecting areas for conservation is unlikely to be as comprehensive or effective as one might wish. Collectively, these insights mean that targeting land management for the protection of species or ecological systems must be far more nuanced and much more challenging than simply setting aside areas of "habitat" and presuming the job is done. The situation is even more challenging for situations where habitat restoration or re-creation is attempted, although for much the same reasons that face habitat protection efforts.

The state of targeting technology

Tools for targeting habitat. Because "habitat" is so important to the conservation and management of species populations, the assessment of habitat requirements for organisms, particularly wildlife, has become a major arena of research activity. Habitat assessments have evolved through several phases: A qualitative phase of describing (qualitatively) habitat associations of species; a quantitative phase of simply correlating measures of individual habitat features with the abundance or density of a species, a preoccupation with increasingly sophisticated (and imponderable) multivariate analyses; and the use of computers to model habitat relationships in a predictive framework (Stauffer 2002). More recently, new statistical approaches, such as classification and regression trees (CART), artificial neural networks, and spatial autocorrelation analyses have brought new power to analyzing habitat patterns, and Akaike's information criterion procedures have provided ways to assess rigorously habitat model performance (Scott et al., 2002; Burnham and Anderson, 1998). Those tools, coupled with the use of remote sensing and GIS, have greatly enhanced our ability to understand how wildlife species respond to and occupy habitat.

At the same time, new approaches and technologies help land managers target areas in which to establish nature reserves. Rather than targeting places for protection based on vague, "pretty places," notions of conservation value, and opportunity, independently of one another, Australian scientists, in particular, advanced computer algorithms for reserve selection based on quantitative features of areas and their contributions to biodiversity protection, contingent on what is already protected ("complementarity"; Margules and Pressey, 2000; Margules, 2005). These approaches and the associated software (e.g., SITES and MARXAN; Groves, 2003) are being used by conservation planners to target and prioritize areas for conservation action based on multiple criteria. At the broad spatial scales of ecoregions (regions

characterized by similar patterns of solar radiation and moisture and, consequently, by similar dominant plants and animals; Bailey, 1998), The Nature Conservancy (TNC) has been conducting ecoregional assessments to identify (target) a subset of areas that collectively represent the overall biological diversity of the ecoregion as a whole. This targeting process incorporates information on the distribution of characteristic and imperiled species, major plant communities, ecosystem types, and patterns of land ownership and management. The areas identified through this targeting process are then the foci for more intensive conservation efforts (e.g., land purchase for protection, conservation easements, cooperative agreements about land use) at more local scales (see Groves, 2003, for a summary of this approach).

Finally, landscape ecology is developing beyond the phase of documenting landscape pattern using GIS and deriving various measures of these patterns using software, such as FRAGSTATS (McGarigal et al., 2002), to incorporate information about how organisms respond to and move through complex landscape mosaics. The dispersal patterns of individuals in a population can be modeled and related to the continuities (e.g., corridors) or discontinuities (e.g., roads, housing developments) in a landscape to predict the probability that seemingly isolated habitats in a landscape may be functionally linked together (Wiens et al., 2002; Wiens, 2001). Formalized approaches of network analysis, diffusion, and percolation are being applied to create neutral models against which the patterns and dynamics of actual landscape linkages may be compared (e.g., Reiners and Driese, 2004; Jongman and Pungetti, 2004; With and King, 1997). These approaches are all aimed at integrating landscape structure with landscape function. How they can be employed to facilitate targeted land management, however, remains to be seen.

Economic instruments for ecosystem targeting. In the case of habitat preservation, most models for selecting sites for reserves to protect species have tended to focus on targeting the best locations for biological reserves to maximize the number of species that can be protected under a given budget constraint or under a restriction on the number or area of sites that can be conserved.

More recently, models have been developed that incorporate spatial criteria in reserve selection to identify the clustering of sites needed to enhance the long-term persistence of species and reduce fragmentation of preserved sites. The few studies that have included cost considerations in targeting habitat reserve sites have shown that budget-constrained site selection can result in more cost-effective conservation, especially where there is considerable spatial heterogeneity in land costs (Polasky et al., 2001). The aim, after all, is not just to be effective in targeting areas or actions for conservation, but to be most effective in applying the resources available to achieve the conservation goals. By applying return-on-investment approaches borrowed from economics to conservation targeting, the biological returns can be balanced against the economic costs, at least in theory (Wilson et al., 2006).

The targeting of sites for protecting species ultimately involves value judgments about whether preserving all species is equally important or if some species are more valuable than others. Often, reserve sites cannot all be protected immediately and excluded sites are threatened by development. The problem of targeting them involves choosing sites through time to include in a network of biological reserves for species conservation. This requires including both expected biodiversity benefits of sites and development risk. Costello and Polasky (2004) and Wilson et al. (2006) pointed out that the timing of targeting efforts is critical. Not only do conservation budgets available up front yield significantly greater biodiversity protection than the same investments delayed until later, but the sequence in which areas are targeted for protection can have a major effect on the cost-effectiveness of an overall conservation program.

Examples of targeted landscape management

Complex, multifaceted concepts like targeted landscape management are often more effectively illustrated through specific examples. Following are several examples that demonstrate new ideas, successes, and different approaches to incorporating targeted management into environmental protection.

Redefining conservation buffers in a landscape context

Water quality protection. Conservation buffers, such as filter strips and riparian forest buffers, can revitalize a host of physical conditions and ecological processes. Buffers can be especially important in extensively cropped landscapes where permanently vegetated areas are scarce. The effectiveness of conservation buffers depends upon the spatial and temporal juxtaposition and interactions between buffer zones and adjacent agricultural areas. For control of nonpoint-source pollution, buffers must be located where they will intercept agricultural runoff; trap a portion of its pollutant load; and stabilize, sequester, and transform those pollutants. General guidelines for federally funded conservation programs target the lower margins of source areas, such as cultivated fields and riparian zones along lake shores, wetlands, and streams. Buffers placed at field margins are closer to the runoff source. Depending upon site conditions, riparian areas can intercept and treat both surface and groundwater runoff.

This configuration of buffer strips woven through an agricultural landscape was recommended in the mid-1990s as an alternative to block configurations for CRP enrollment. But there are finer scale spatial patterns in agricultural landscapes that largely determine the water quality impact of buffer installations. Sources, transport pathways, and buffering suitabilities vary across landscapes, reflecting both inherent site properties and management influences. For example, source-area contributions vary with soil and slope properties of fields, as well as with tillage and other land treatments (e.g., Wischmeier and Smith, 1978). Overland flow is far from uniform, flowing into topographic swales, being diverted into drainage tiles and ditches, and even steered by the unevenness created by tillage furrows and ridges (Bren, 1998; Dosskey et al., 2003; Souchere et al., 1998; Tomer et al., 2003). Buffer effectiveness also varies with slope, soil type, groundwater depth, and subsurface geomorphology. In general, buffer capabilities may be greater in the upper reaches of watersheds than along the main stems of streams (Burkart et al., 2004). Buffer interactions with sources and pathways also occur, as pollutant-trapping efficiency largely varies with the amount and timing of intercepted runoff (Dosskey et al., 2002). Consequently, opportunities to effect watershed water quality are greater at some locations in a watershed than others. Targeting buffers to locations where impacts are likely to be greatest and avoiding those where impacts are likely to be small promises to improve substantially the conservation efficiency of buffers at the watershed or landscape scale.

Methods have been developed for assessing landscape patterns and identifying critical buffer locations. Ranking source areas by analysis of inherent site conditions and management, for example, by using USLE- or CREAMS-based models, has long been a focus of water quality improvement projects. Recent approaches have been developed that identify overland flow pathways and buffer suitability patterns. Terrain analysis using digital elevation models can identify probable patterns of runoff pathways and, at the stream-reach scale, relative baseflow contributions to streams (Tomer et al., 2003; Burkart et al., 2004). Topographic analysis also helps predict where variable-source areas, which generate overland flow nearest the stream, will occur (e.g., Lyon et al., 2004). Upslope contributing area and local slope of riparian areas are the most critical topographic parameters for making these determinations. Soil surveys can be used to identify the best-suited locations for buffer filtering of surface runoff using a model that accounts for both potential source size and buffer capacity (Dosskey et al., 2006). Soil surveys can also be used to identify locations where shallow groundwater occurs, which may be treated by buffers (Dosskey et al., 2006; Gold et al., 2001; Rosenblatt et al., 2001). Regional geomorphic patterns have been interpreted to assess the potential for riparian buffers to impact runoff and baseflow water quality (Lowrance et al., 1997). Landscape-scale geomorphic patterns may also be interpreted for potential nitrate attenuation in riparian groundwater (Vidon and Hill, 2006). At an even finer scale, a precision approach to buffer site design has been proposed that combines a spatial assessment of source-area loads, runoff pathways, and buffering capacities to recommend appropriate and varying buffer widths across landscapes (Dosskey et al., 2005). These methods all aim to identify locations in a landscape where buffer impact on water quality is likely to be greater than in others.

A buffer in an agricultural landscape for non-

point-source pollution control that is based on recent science is likely to look somewhat different than the buffer strip model of the 1990s (Figure 1). Buffers should be most consistently located along the upper reaches of the watershed (Burkart et al., 2004). Wider buffers should be installed where surface runoff loads will be greatest or variable-source-area hydrology is most active (e.g., Walter et al., 2005; Qui, 2003). Riparian buffers need not be continuous or of constant width, but concentrated in areas where surface runoff is most readily intercepted and filtered, where saturation overland flow is likely to be generated from variable-source areas, and where buffer vegetation may have the greatest opportunity to influence shallow groundwater. There may be opportunities to combine topographic analyses and soil survey information in a complementary fashion to further enhance targeting of conservation buffers (Tomer et al., 2006).

The gains in conservation efficiency expected by targeting water quality buffers will be moderated somewhat by issues that include stratigraphic and pedogenic controls on subsurface flow patterns (Kung, 1990; Lowrance et al., 1997; Simpkins et al., 2002; Agier et al., 2005) and effects of hydrologic modifications, particularly artificial drainage (Wu and Babcock, 1999; Goehring et al., 2001; Dosskey et al., 2003). One solution to address artificial drainage is to incorporate constructed wetlands into riparian buffers that intercept and treat subsurface drainage water before it reaches streams (Schultz et al., 1995). Alterations of topographic and hydrologic patterns in intensively cultivated landscapes will probably complicate the use of existing assessment methods. For targeting techniques to overcome these problems, detailed and recent landscape data may be required for field-scale planning.

Habitat. Conservation buffers can create habitat that bolsters wildlife populations in agricultural landscapes. There is often too little suitable habitat, especially where intensive agricultural activities and infrastructure occupy most of the landscape. Remnants of native habitat may be too small to support viable wildlife populations. Isolation of habitat fragments restricts or prevents daily movements between sources of food, water, and cover; annual migrations to reproduction areas; and movement from remnants to newly

A   **B**

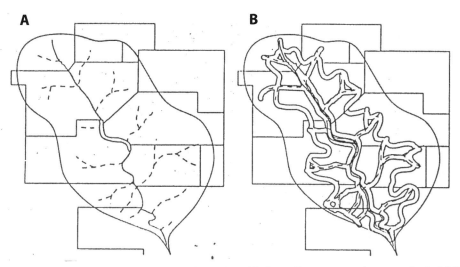

Figure 1. A conceptual diagram comparing (A) a traditional CRP block enrollment pattern in a watershed with (B) an area of land similar to a block, but set aside in a configuration of buffer strips (as shown in National Research Council, 1993). This diagram illustrates a recommended shift in approach to land retirement from blocks to targeted strips in order to increase the conservation effectiveness of CRP and related programs.

planted habitat areas. Buffers can help bolster viable wildlife populations by increasing the total area of suitable habitat, reducing the isolation of habitat remnants, and providing a management focus for improving habitat quality.

Greater wildlife impacts can be obtained from conservation buffers by appropriately locating them within agricultural landscapes. Corridors are a particularly important configuration in fragmented landscapes for increasing the viability of small, isolated patches; providing access to food, water, cover and other critical needs that may exist in different parts of the landscape; and creating avenues for dispersal and repopulating new habitat areas (U.S. Department of Agriculture-Natural Resources Conservation Service, 1999). Riparian areas offer disproportionately higher overall habitat value than upland areas (Naiman et al., 1993), in part because of their intrinsically greater habitat quality and their continuity through landscapes. Using conservation buffers to close existing gaps between habitat patches and create continuous riparian corridors can enhance conservation efficiency for terrestrial wildlife. GIS, coupled with land use and land cover maps, is a particularly helpful tool for determining critical locations (e.g., Bentrup and Kellerman, 2004)

Riparian areas are also major habitat elements for fish assemblages in streams and riv-

ers (Schlosser, 1991). Adjacent vegetation affects channel form and structural diversity by contributing large organic debris to channels and by its influence on bank erosion and sediment deposition patterns (Trimble, 2004). Substantial inputs of organic matter and nutrients from riparian zones fuel the aquatic food chain (Vannote et al., 1980).

Opportunities for conservation buffers to benefit aquatic resources are generally greater in the upper reaches of stream systems, where interactions between riparian and aquatic systems are greatest (National Research Council, 2002). Shading effects on water temperature and accumulations of large woody debris are greater in smaller streams (Harmon et al., 1986; Hewlett and Fortson, 1982; Karr and Schlosser, 1978). Smaller streams also derive a greater proportion of their fish food resources from riparian areas (Newbold et al., 1980) and comprise a much greater collective length of streams in watersheds.

The conservation efficiency of buffers can be enhanced further by integrating placement criteria to benefit multiple resources. Riparian areas and corridors that connect them to remnant upland patches are particularly valuable for creating terrestrial habitat. Riparian areas in the upper reaches of watersheds are particularly influential on aquatic habitat and nonpoint-source pollution. Finer scale targeting, using topographic and soils information, should enhance the effectiveness of conservation buffers for improving water quality.

Hydrologically sensitive areas

One of the many new perspectives to come out of the New York City watershed protection program (Walter and Walter, 1999) is the concept of hydrologically sensitive areas—those areas especially prone to generating runoff (New York City Department of Environmental Protection, 1991; Walter et al., 2000, 2001; Agnew et al., 2006). The hydrologically-sensitive-areas management concept is simply to avoid potentially polluting activities (e.g., manure spreading) on areas most prone to generating runoff or to target water quality protection strategies on those areas that are both hydrologically sensitive and potential nutrient- or pollutant-loading areas. Some hydrologocially sensitive areas, such as paved barnyards, are obvious. Throughout most of the northeastern United States, however, runoff is generated from areas that become saturated. These areas can be very dynamic, expanding during wet periods and shrinking or disappearing entirely during dry periods; these dynamic areas are referred to as variable-source areas (e.g., Hewlett and Hibbert, 1967; Dunne and Black, 1970a,b) (Figure 2a). Research efforts as part of the New York City watershed program demonstrated that distributed models could accurately locate variable-source areas (Frankenberger et al., 1999; Mehta et al., 2004; Gerard-Merchant et al., 2006) and that by adopting associated targeted BMPs, phosphorus loading to streams could be reduced (Walter et al., 2001; Bishop et al., 2005; Hively et al., 2006). However, linking the basic hydrologic science describing where and when these areas will appear with policy and management structures has proven the most vexing challenge (Gburek et al., 2002), in part because hydrologically sensitive areas are both spatially and temporally distributed (Figure 2b).

Initial attempts to develop hydrologically-sensitive-area guidelines were incorporated into the New York phosphorus index (Geohring et al., 2002; Czymmek et al., 2003) by using a combination of soils information (e.g., flooding frequency) and set-back distances from identifiable runoff flowpaths (e.g., ephemeral streams). Even these initial achievements were a marked change from the original runoff-risk criteria

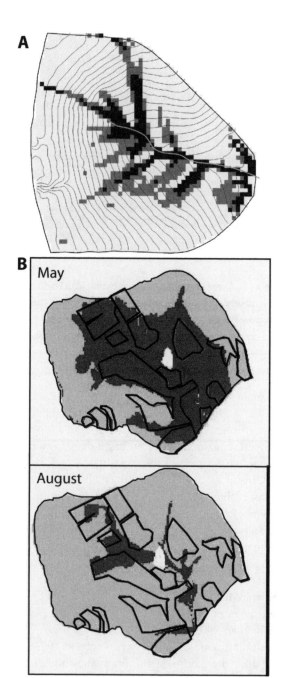

Figure 2. Examples of how variable source (runoff generating) areas (A) are distributed across the landscape (from Tomer, 2004) and (B) vary over the course of a year (from Agnew et al., 2006). In A, dark-gray and medium-gray areas represent frequently and occasionally runoff generating areas, respectively. In B, darker gray areas represent areas generating runoff at least five days in a month and the outlined areas delineate fields.

proposed for phosphorus indices, which cor-related steep slopes and low infiltration capaci-ties with high runoff potential (U.S. Department of Agriculture-Natural Resources Conservation Service, 1994). Neither of these factors correlates well with field observations of runoff-generat-ing areas in the northeastern United States (e.g., Dunne and Black, 1970a,b; Frankenberger et al., 1999; Walter et al., 2003; Mehta et al., 2004; Easton et al., 2006). Interestingly, 84 percent of the cur-rent phosphorus indices have adopted hydrologi-cally-sensitive-area-type criteria for identifying areas of high runoff risk (Gburek et al., 2006). One challenge in defining an hydrologically sensitive area is specifying a risk level, although research-ers have demonstrated that levels can be quanti-fied by distributed modeling (e.g., Walter et al., 2000, 2001), field measurements (e.g., Lyon et al., 2006a,c), and remote sensing (e.g., Verhoest et al., 1998). Several efforts are underway to iden-tify and quantify hydrologically sensitive areas using GIS (e.g., Agnew et al., 2006), including internet-based tools, such as Google Earth (Lyon et al., 2006b). Additionally, local agencies (e.g., Tompkins County, New York Department of Environmental Planning) have begun requesting GIS data for their counties, in part as a means for delineating riparian buffers (Walter et al., 2005). These tools provide planners with information about how hydrologically-sensitive-area risk is distributed across the landscape and throughout the year. Consistency of data in terms of source and quality will remain a challenge to devising targeting tools based on hydrologically-sensitive-area assessments that can be consistently applied from a policy standpoint. Such tools will require a certain level of professional judgment in imple-menting targeted management in hydrologically sensitive areas.

Targeting criteria based on soil-test phosphorus

High livestock densities and associated land-applied manure pose serious threats to water quality across the United States and Canada (Sharpley et al., 1998; Canada-Alberta Environ-mentally Sustainable Agriculture Agreement Water Quality Committee, 1998). Livestock manure is typically applied to cropland based on crop nitrogen requirements. However, because the nitrogen-to-phosphorus ratio in manure is substantially lower than that of crops, accumula-tion of phosphorus on manured cropland is com-mon (Sharpley et al., 1998). Surface runoff from cropland high in soil phosphorus can be a pri-mary source of phosphorus enrichment of surface waters and subsequent eutrophication risks.

In Alberta, Canada, planners are developing targeted manure management that limits appli-cation based on site-specific soil-test phosphorus levels (Jedrych et al., 2006). Targeting was at the soil polygon scale because this scale represents the most detailed level of available soils informa-tion. Allowable soil-test phosphorus levels (modi-fied Kelowna method) were quantified using the Water Erosion Prediction Project (WEPP) model to predict surface runoff, assuming water quality objectives of 0.5 milligram per liter and 1.0 mil-ligram per liter and employing empirical relation-ships between total phosphorus concentrations in runoff and soil-test phosphorus in the top 15 centimeters (6 inches) of soil (Little et al., 2006). This method allows higher soil-test phosphorus values and higher associated total phosphorus in runoff for fields with lower runoff potential, and visa versa for fields with higher runoff potential. The spatial distribution of soil-test phosphorus was markedly different between the strict and more relaxed water quality criteria, 0.5 milligram per liter and 1.0 milligram per liter, respectively (Jedrych et al., 2006) (Figure 3). Perhaps most notably, soil-test phosphorus limits of less than 60 milligrams per kilogram (60 parts per million) were required for more than 80 percent of the agricultural land base in the province to meet the 0.5 milligram per liter water quality objective and dropped to less than 50 percent for the 1.0 milli-gram per liter objective (Figure 3).

The main advantage of this soil-test phospho-rus targeting approach is that it uses the rela-tive runoff potential, in this case determined by WEPP, instead of actual runoff values to calcu-late soil-test phosphorus limits; this eliminates the need for expensive WEPP model calibration. Unfortunately, because runoff estimates may be inaccurate, allowable soil-test phosphorus limits are similarly inaccurate. This approach, therefore, provides only a relative ranking of targeted prior-ity areas.

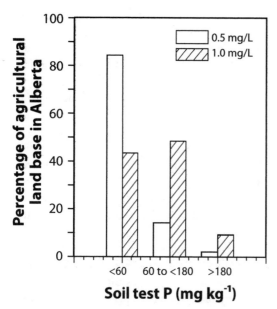

Figure 3. Soil test phosphorus limits for agricultural land base in Alberta required to meet total phosphorus water quality objectives of 0.5 and 1.0 milligram per liter.

Targeting places for biodiversity conservation

Recognizing that the needs for managing or protecting places for biodiversity conservation far outstrip the resources available to achieve that management or protection, several conservation organizations have used various approaches to prioritize (i.e., target) areas for concerted action, at scales from local to global. Rather than reviewing targeting approaches at the local or regional scales discussed in the previous examples, we briefly consider some efforts focused at a global scale. Groves (2003) provided an excellent review of targeting land management for conservation (a.k.a., conservation planning) at local and regional scales.

At a global scale, Conservation International has identified a number of "hotspots" of biodiversity that, if adequately protected, would contribute to preserving a substantial portion of the earth's species (Myers et al., 2000; Mittermeier et al., 1998). These areas are targeted largely on the basis of plant species richness (and secondarily vertebrates) and vulnerability. Not surprisingly, nearly all of the hotspots are in tropical and sub-tropical regions (Figure 4).

The World Wildlife Fund (WWF) has used species occurrences, the presence of distinct ecosystems and ecological processes, and the biological distinctiveness of areas to target a more widely distributed set of areas, the "Global 200 Ecoregions," for conservation action (Figure 4) (Olson and Dinerstein, 1998).

To address conservation priorities globally, TNC has developed an even more comprehensive, data-driven approach to targeting ecoregions within major habitat types of the earth (e.g., temperate conifer forests, deserts, and xeric shrublands), partitioned among biogeographic realms (e.g., Nearctic, Australasia). The approach emphasizes those ecoregions in which conservation actions may make the greatest contribution to the benchmark goal of effective conservation or management of 10 percent of a habitat type within a realm. Like the WWF Global 200, this approach is aimed at ensuring adequate conservation of representative biodiversity everywhere on earth – the "coldspots" as well as the hotspots (Kareiva and Marvier, 2003). Still other organizations have used different criteria to target areas for conservation at a global scale (Figure 4). Collectively, some areas are targeted by everyone, but others emerge only when representation is emphasized in the targeting process (e.g., Brooks et al., 2006).

All of these global targeting approaches emphasize in one way or another the spatial distribution of biological diversity and threats to that diversity at very broad spatial scales. TNC has taken an important step toward broadening the targeting criteria by using country-level data on political governance, economic risks, social well-being, and civil society development to define "enabling conditions." Not all places in which conservation is a high priority on biological grounds are conducive to investing in conservation or conducting work on the ground for sociopolitical or economic reasons. Moreover, conservation in most parts of the world cannot be achieved through strict protection of parks and reserves (although this certainly helps). Conservation must also be done in places where people live and work – people and their well-being must be part of the conservation equation and thus part of targeting land management (Miller and Hobbs, 2002; Millennium Ecosystem Assessment, 2005). One way to include social, political, and economic factors in conservation targeting is by using return-on-investment tools

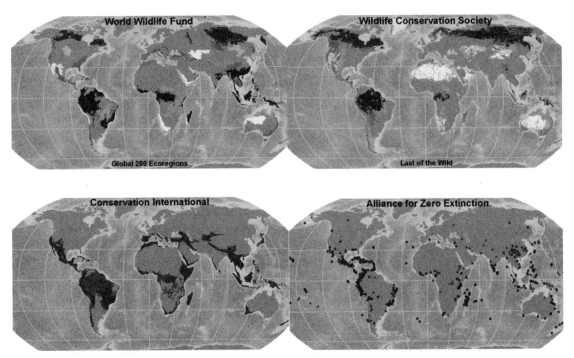

Source: Maps courtesy of the Nature Conservancy.

Figure 4. Areas targeted for biodiversity protection by four conservation organizations using different global targeting criteria.

and models from economics to prioritize areas for conservation or the actions that might be taken to achieve conservation most economically in those areas. While this work is still in an exploratory phase (e.g., Wilson et al., 2006; Polasky et al., 2001), it already has shown that traditional (even sophisticated) ways of targeting places based on their apparent biodiversity value alone may not lead to the most effective use of limited funds. Combined with new methods of evaluating the success of conservation actions (e.g., Ferraro and Pattanayak, 2006), it may be possible to target conservation to achieve the greatest "biobang for the buck." Such prioritizations must include socioeconomic and political factors as well as the biodiversity value of targeted places.

Challenges and future directions

We mentioned previously the substantial progress that has been made toward developing rigorous and comprehensive conservation planning (targeting) at multiple scales. Yet the greater understanding of habitats and habitat relation-

ships of organisms across those scales has also revealed several bothersome realities that raise doubts about how much we really do understand ecological systems. These realities, in turn, affect our ability to target management actions in ways that will actually produce desired outcomes. We note here two of these realities.

First, the elements of ecological systems rarely act in isolation, and management activities rarely have only a single consequence. There are interactions and interdependencies to consider. Sometimes these are complementary, for example, when habitat protection for one endangered species results in habitat protection for a suite of associated species, or when measures to enhance seasonal streamflow for fish migration also result in a flushing of sediment from basins and improved water quality and biotic integrity. In other situations, however, land management targeted by one set of criteria may have unintended consequences or negative impacts on other components of an ecological system. For example, use of prescribed burning to maintain an open understory in woodlands may favor hawks that prey on ground-dwelling mammals, but negatively affect not only mammal populations, but also bird species that require an intermediate shrub layer for nesting. Preventing human intrusion into a forested watershed that serves as a municipal water supply may protect water quality, but it may also preclude human recreation in the watershed, leading to reduced support for watershed conservation efforts. Conservation is increasingly about tradeoffs among competing interests or targets. Return-on-investment or cost-benefit analyses may provide pathways to quantifying or even optimizing these tradeoffs, but this requires knowledge of what to include in the analysis, how to weight the competing interests, and what metrics to use as measures of "return" and "investment." In the world of conservation, these metrics are often noneconomic, but the science of incorporating noneconomic factors into what are essentially economic models is not yet well developed.

The second reality has to do with "thresholds"–"tipping points" in contemporary parlance (Gladwell, 2002). Targeted land management is often conducted as if the targeted systems and the outcomes of targeted actions were stable, or at least varying within well-defined and prescribed limits. But ecological systems are dynamic in time and space. Not only that, they are inconsistently dynamic. Natural or human-induced changes in system properties that appear to be gradual and continuous may sometimes lead to sudden and irreversible shifts to some other system or set of prevailing conditions, as envisioned in so-called "state-and-transition" models (e.g., Bestelmeyer et al., 2003). Examples of such changes in arid lands, as from grasslands to shrublands or vice versa, are legion (e.g., Bestelmeyer, 2006), but they occur in virtually every type of terrestrial and aquatic habitat (e.g., Groffman et al., 2006). Sudden, irreversible changes resulting from global climate change are of increasing concern (e.g., Lovejoy and Hannah, 2005). The challenge to targeted land management is clear: Not only must we recognize the appropriate targets for our management actions to have the intended effects, but we are managing with reference to a moving target and (to stretch the metaphor) one that may change shape, identity, and context suddenly and without warning. A better predictive science of ecological thresholds would enhance targeted land management.

Summary and conclusions

In conclusion, we emphasize four points. First, although conservation practices have always had some degree of "targeting," targeting approaches have become increasingly refined with our improved understanding of how individual parts of the landscape are interconnected. Concurrently, technological advances in GIS, remote sensing, and other methods of gathering, compiling, and analyzing geospatial data have facilitated the linkage between science and the application of targeted land and water management for conservation. But there is a catch. Although the scientific bases for targeted protection are generally good and getting better, the ease with which GIS and other technologies can be used can facilitate the inappropriate use of older data and models. These may have been effective in meeting the objectives of their day, but they do not incorporate recent scientific advances that have created, for example, data with a high degree of spatial resolution at multiple scales or analytical procedures that can incorporate the nonlinear or threshold dynamics of ecological systems. As ever-more-powerful technological and analytical tools become available, it is important to keep in mind that such tools are only as good as the information they have to work with. Assumptions about the utility of data must be clearly stated. Great tools cannot produce great targeting with inadequate data, or with hidden assumptions about oversold data.

Second, determining how and what to target, with what degree of precision, should be driven by one's objectives. The targeting approach that one takes to evaluate the consequences of agricultural practices on stream water quality in the Midwest will likely differ from that applied to a similar situation in the Intermountain West, or in Guatemala or Bangladesh. The tools used to target an area for the application of soil erosion mitigation will likely differ from those used to target wildlife habitat conservation in the same area. There is no great insight here. But what this does highlight is the critical importance of clearly stating the goals and objectives of a conservation project, in operational terms, at the outset of any targeting effort. And the objectives must align with the characteristics of the biophysical system and the management strategies to be applied to that system (Figure 5).

Third, issues of scale are of overriding importance. The challenges to water quality conservation are not the same at the scale of a local farm, in which the primary concern may be with surface and subsurface runoff from manure spreading on fields into a small stream, as at the scale of a large regional watershed (e.g., the Upper Mississippi River), where a host of point and nonpoint sources act to influence water quality. Scientific work over the past two decades has shown the many ways in which both ecological patterns and processes and our perceptions of these patterns and processes are influenced by the scale of reference (e.g., Wiens, 1989; Peterson and Parker, 1998), although this work has not yet produced a workable way of predicting these effects. What is clear, however, is that targeting efforts must be scale-sensitive. This means that targeting must focus at scales where the scales of the system, management practices, and objectives coincide (Figure 5). Indeed, a goal of targeting should be to move these components into greater scale concordance (i.e., greater overlap in the Venn diagram of Figure 5).

Finally, our emphasis in this chapter has been on the science that underlies efforts to target conservation efforts to enhance the effectiveness (and cost-effectiveness) of management, mitigation, and restoration. But science alone does not have all the answers. Science can inform one about where conservation efforts might do the most good, when they would best be applied, and how to gauge whether or not they are working as desired. But whether or not the targeting recommended by science can or should be implemented involves other considerations. Land and water management are conducted in an economic, social, and political context, and this context is what determines whether the science-based course of action is in fact feasible or desirable. The mitigation or restoration of riparian areas as buffers against runoff from agricultural land, for example, may be too expensive for the benefits to be derived, or the lands identified through a habitat targeting procedure for wildlife protection may be unavailable. And we have already mentioned the bothersome issue of tradeoffs – a conservation action that is highly desired by some segments of society may be vigorously opposed by other segments. If compromises are possible, one must then consider whether the diminished

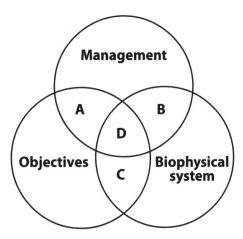

Figure 5. The intersection of objectives with management actions and the biophysical system under consideration. Targeting can aim to align objectives with management actions (A), management with the systems (B), or objectives with the biophysical system (C). Ideally, targeting should aim to align all three components simultaneously (D). Because objectives, management, and the biophysical system may be expressed or operate at different scales, targeting should also aim to align all three components within a common scale of reference (D).

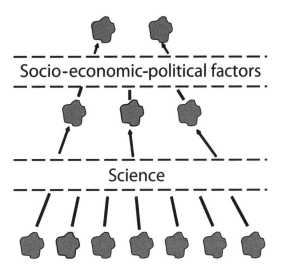

Figure 6. The role of science and socio-economic-political factors in targeting places for conservation. A set of possible locations for conservation action (bottom) is filtered by science-based targeting to identify a subset that will most effectively yield conservation benefits; socio-economic-political considerations then determine in which elements of this subset conservation efforts may be feasible or acceptable.

benefits make the actions worth the effort. These realities should not be taken as weakening the importance of science in targeting, however. In our view, science provides the foundation, the first-pass assessment of what should be done, based on a rigorous analysis of data from multiple sources. Once that judgment has been made, then the socioeconomic and political realities should be incorporated to determine which of the possible targeting actions identified by scientific analysis actually merits implementation (Figure 6).

References

Agnew, L.J., S. Lyon, P. Gérard-Marchant, V.B. Collins, A.J. Lembo, T.S. Steenhuis, and M.T. Walter. 2006. Identifying hydrologically sensitive areas: Bridging science and application. *Journal of Environmental Management.* 78: 64-76.

Angier, J.T., G.W. McCarty, and K.L. Prestegaard. 2005. Hydrology of a first order riparian zone and stream, mid-Atlantic coastal plain, Maryland. *Journal of Hydrology* 308:149-166.

Arnold, J.G., P.M. Allen, and G. Bernhardt. 1993. A comprehensive surface-groundwater flow model. *Journal of Hydrology* 142:47-69.

Babcock B.A., P.G. Lakshminarayan, J. Wu, and D. Zilberman. 1996. The economics of a public fund for environmental amenities: A study of CRP contracts. *American Journal of Agricultural Economics* 78(4): 961-71.

Babcock B.A., P.G. Lakshminarayan, J. Wu, and D. Zilberman. 1997. Targeting Tools for the Purchase of Environmental Amenities. *Land Economics* 73(3): 325-39.

Bailey, R.G. 1998. Ecoregions: The ecosystem geography of the oceans and continents. Springer, New York, New York.

Batie, S.S., and D.E. Ervin. 1999. Flexible incentives for environmental management in agriculture: A typology. In: F.Casey, A. Schmitz, S.Swinton, and D. Zilberman, editiors, Flexible Incentives for the Adoption of Environmental Technologies in Agriculture. Kluwer Academic Publishers, Norwell, Massachusetts. Pp. 55-78.

Benbrook, C.M. 1988. First principles: The definition of highly erodible land and tolerable soil loss. *Journal of Soil and water Conservation* 43:35-38.

Bentrup, G., and T. Kellerman. 2004. Where should buffers go? Modeling riparian habitat connectivity in northeast Kansas. *Journal of Soil and Water Conservation* 59:209-215.

Bestelmeyer, B.T. 2006. Threshold concepts and their use in rangeland management and restoration: The good, the bad, and the insidious. *Restoration Ecology* 14: 325-329.

Bestelmeyer, B.T., J.R. Brown, K.M. Havstad, G. Chavez, R. Alexander, and J.E. Herrick. 2003. Development and use of state-and-transition models for rangelands. *Journal of Range Management* 56: 114-126.

Beven, K., and P. Germann. 1982. Macropores and water flow in soils. *Water Resources Research* 18:1311-1325.

Bishop P.L., W.D. Hively, J.R.Stedinger, M.R. Rafferty, J.L. Lojpersberger, and J.A. Bloomfield. 2005. Multivariate analysis of paired watershed data to evaluate agricultural best management practice effects on stream water phosphorus. *Journal of Environmental Quality* 34 (3): 1087-1101.

Bren, L.J. 1998. The geometry of a constant-loading design method for humid watersheds. *Forest Ecology and Management* 110:113-125.

Brooks, T.M., R.A. Mittermeier, G.A.B. da Fonseca, J. Gerlach, M. Hoffmann, J.F. Lamoreux, C.G. Mittermeier, J.D. Pilgrim, and A.S.L. Rodrigues. 2006. Global biodiversity conservation priorities. *Science* 313: 58-61.

Browne, B.A. and N.M. Guldan. 2005. Understanding long-term baseflow water quality trends using a synoptic survey of the ground water-surface water interface, central Wisconsin. *Journal of Environmental Quality* 34:825-835.

Burkart, M.R., D.E. James, and M.D. Tomer. 2004. Hydrologic and terrain variables to aid strategic location of riparian buffers. *Journal of Soil and Water Conservation* 59:216-223.

Burnham, K.P. and D.R. Anderson. 1998. *Model selection and inference: A practical information-theoretic approach.* New York: Springer.

Busenberg, E. and L.N. Plummer. 2000. Dating young groundwater with sulfur hexafluoride: natural and anthropogenic sources of sulfur hexafluoride. *Water Resources Research* 26(10): 3011-3030.

Canada-Alberta Environmentally Sustainable Agriculture Agreement Water Quality Committee. 1998. Agricultural impacts on water quality in Alberta-an initial assessment. Alberta Agriculture, Food and Rural Development, Edmonton, Alberta.

Carlyle G.C. and A.R. Hill. 2001. Groundwater phosphate dynamics in a river riparian zone: effects of hydrologic flowpaths, lithology and redox chemistry. *Journal of Hydrology* 247(3-4): 151-168.

Carson, A.C., B.L. Shear, M.R. Ellersleck, and A.S. Asfaw. 2001. Identification of of fecal Escherichia coli from humans and animals by ribotyping. *Applied Environmental Microbiology* 67: 1503-1507.

Christophersen, N., C. Neal, R.D. Hooper, R.D. Vogt, and S. Andersen. 1990. Modelling stream water chemistry as a mixture of soilwater end members – a step towards second generation acidification models. *Journal of Hydrology* 116:307-320.

Collins, A.L., and D.E. Walling. 2002. Selecting fingerprint properties for discriminating potential suspended sediment sources in river basins. *Journal of Hydrology* 261:218-244.

Costello, C. and S. Polasky. 2004. Dynamic reserve site selection, *Resource and Energy Economics, Special Issue* 26(2):157-174

Czymmek, K.J., Q.M. Ketterings, L.D. Geohring, and G.L. Albrecht. 2003. The New York phosphorus runoff index user's manual and documentation. Cornell University, Ithaca, New York. 64 pp. (//nmsp.css.cornell.edu/publications/pindex.asp).

Dale, V.H., and R.A. Haeuber, editors. 2001. Applying ecological principles to land management. Springer, New York, New York.

Dietz, M.E., J.C. Clausen, and K.K. Filchak. 2004. Education and changes in residential nonpoint source pollution. *Environmental Management* 34(5): 684-690.

Dillaha T.A., J.H. Sherrard, D. Lee, S. Mostaghimi, and V.O. Shanholtz. 1988. Evaluation of vegetative filter strips as a best management practice for feed lots. *Journal of the Water Pollution. Control Federation* 60(7): 1231-1238.

Dinnes, D.L., D.L. Karlen, D.B. Jaynes, T.C. Kaspar, J.L. Hatfield, T.S. Colvin, and C.A. Cambardella. 2002. Nirogen management strategies to reduce nitrate leaching in tile-drained Midwest soils. *Agronomy Journal* 94(1):153-171.

Dombek, P.E., L.K. Johnson, S.T. Zimmerly, and M.J. Sadowsky. 2000. Use of repetitive DNA sequences and the PCR to differentiate Escherichia coli isolates from human and animal sources. *Applied Environmental Microbiology* 66:2572-2577.

Dosskey, M.G., D.E. Eisenhauer, and M.J. Helmers. 2005. Establishing conservation buffers using precision information. *Journal of Soil and Water Conservation* 60:349-354.

Dosskey, M.G., M.J. Helmers, D.E. Eisenhauer, T.G. Franti, and K.D. Hoagland. 2002. Assessment of concentrated flow through riparian buffers. *Journal of Soil and Water Conservation* 57:336-343.

Dosskey, M.G., M.J. Helmers, and D.E. Eisenhauer. 2006. An approach for using soil surveys to guide the placement of water quality buffers. *Journal of Soil and Water Conservation* 61(6):344-354.

Dosskey, M., M. Helmers, D. Eisenhauer, T. Franti, and K. Hoagland. 2003. Hydrologic routing of farm runoff and implications for riparian buffers. In J.D. Williams and D. Kolpin, editors, Agricultural Hydrology and Water Quality. American Water Resources Association, Middleburg, Virginia. (CD-ROM)

Dramstad, W.E., J.D. Olson, and R.T.T. Forman. 1996. Landscape ecology principles in landscape architecture and land-use planning. Island Press, Washington, D.C. 80 pp.

Dunne, T., and R.D. Black. 1970a. An experimental investigation of runoff production in permeable soils. *Water Resources Research* 6(2): 478-490.

Dunne, T., and R.D. Black. 1970b. Partial area contributions to storm runoff in a small New-England watershed. *Water Resources Research* 6(5): 1,296-1,311.

Easton, Z.M., P. Gérard-Marchant, M.T. Walter, A.M. Petrovic, and T.S. Steenhuis. 2006. Hydrologic assessment of an urban variable source watershed in the Northeast United States. *Water Resources Research* (in press).

Edwards, D.R., T.C. Daniel, and O. Marbun. 1992. Determination of best timing of poultry waste disposal: A modeling approach. *Water Resources Bulletin* 28(3):487-494.

Ferraro, P.J. and S.K. Pattanayak. 2006. Money for nothing? A call for empirical evaluation of biodiversity conservation investments. *PLoS Biology* 4(4): 482-488

Forman, R.T.T., and M. Godron. 1986. *Landscape Ecology*. John Wiley and Sons, New York, 619 p.

Flury, M., H. Flühler, W.A. Jury, and J. Leuenberger. 1994. Susceptibility of soils to preferential flow of water: A field study. *Water Resources Research*. 30(7):1945-1954.

Frankenberger, J.R., E.S. Brooks, M.T. Walter, M.F. Walter, and T.S. Steenhuis. 1999. A GIS-based variable source area model. *Hydrological Processes* 13(6): 804-822.

Gardner, R.H., W.M. Kemp, V.S. Kennedy, and J.E. Petersen, editors. 2001. Scaling relations in experimental ecology. Columbia University Press, New York, New York.

Gatcher, R., J.M. Ngatiah, and C. Stamm. 1998. Transport of phosphate from soil to surface waters by preferential flow. *Environmental Science and Technology* 32(13):1,865-1,869.

Garen, D.C., and D.S. Moore. 2005. Curve number hydrology in water quality modeling: Uses, abuses, and future directions. *Journal of the American Water Resources Association* 41(2):377-388.

Gburek, W.J., C.C. Drungil, M.S. Srinivasan, B.A. Needelman, and D.E. Woodward. 2002. Variable-source-area controls on phosphorus transport: Bridging the gap between research and design. *Journal of Soil and Water Conservation* 57(6): 534-543.

Gburek, W.J., M.T. Walter, and T.S. Steenhuis. 2006. Impact of real watershed hydrology on modeling phosphorus transport. Presentation and Abstract, Sera17 Modeling Phosphorus Transport in Agroecosystems: Joining Users, Developers, and Scientists. Cornell University July 31-August 2, 2006.

Genereux, D.P., and R.P. Hooper. 1998. Oxygen and hydrogen isotopes in rainfall-runoff studies. p. 319-346 In C. Kendall and J.J. McConnell, editors, Isotope tracers in catchment hydrology. Elsevier Science, Amsterdam, The Netherlands.

Geohring, L.D. O.V. McHugh, M.T. Walter, T.S. Steenhuis, M.S. Akthar, and M.F. Walter. 2001. Phosphorus transport into subsurface drains by macropores after manure applications: Implications for best manure management practices. Soil Science 166(12):896-909.

Geohring, L.D., T.S. Steenhuis, M.T. Walter, M.F. Walter, Q.M. Ketterings, and K.J. Czymmek. 2002. Phosphorus risk assessment tools for New York State. ASAE-CIGR Paper 0022071. American Society of Agricultural Engineers Annual International Meeting / Commission Internationale du Génie Rural XVth World Congress, July 28-31, Chicago, IL.

Gerard-Marchant P., W.D. Hively, and T.S. Steenhuis. 2006. Distributed hydrological modelling of total dissolved phosphorus transport in an agricultural landscape, part I: distributed runoff generation. *Hydrology and Earth Systems Sciences* 10(2): 245-261.

Gessler, P.E., O.A. Chadwick, F. Chamran, L. Althouse, and K. Holmes. 2000. Modeling soil-landscape and ecosystem properties using terrain attributes. *Soil Science Society of America Journal* 64:2046-2056.

Gladwell, M. 2002. The tipping point: How little things can make a big difference. Little, Brown and Company, New York, New York.

Gold, A.J., P.M. Groffman, K. Addy, D.Q. Kellogg, M. Stolt, A.E. Rosenblatt. 2001. Landscape attributes as controls on ground water nitrate removal capacity of riparian zones. *Journal of the American Water Resources Association* 37:1,457-1,464.

Groffman, P.M., N.J. Boulware, W.C. Zipperer, R.V. Pouyat, L.E. Band, and M.F. Colosimo. 2002. Soil nitrogen cycle processes in urban Riparian zones. *Environmental Science & Technology* 36(21): 4,547-4,552.

Groffman P.M. and M.K. Crawford. 2003. Denitrification potential in urban riparian zones. *Journal of Environmental Quality* 32 (3): 1,144-1,149.

Groffman, P.M., J.S. Baron, T. Blett, A.J. Gold, I. Goodman, L.H. Gunderson, B.M. Levinson, M.A. Palmer, H.W. Paerl, G.D. Peterson, N.L. Poff, D.W. Rejeski, J.F. Reynolds, M.G. Turner, K.C. Weathers, and J. Wiens. 2006. Ecological thresholds: The key to successful environmental management or an important concept with no practical application? *Ecosystems* 9: 1-13.

Groves, C.R. 2003. Drafting a conservation blueprint. Island Press, Washington D.C.

Haith, D.A., and L.L. Shoemaker. 1987. Generalized watershed loading functions for strem-flow nutrients. *Water Resources Research* 23(3):471-478.

Harmon, M.E., J.F. Franklin, F.J. Swanson, P. Sollins, S.V. Gregory, J.D. Lattin, N.H. Anderson, S.P. Kline, N.G. Aumen, J.R. Sedell, G.W. Lienkaemper, K. Cromack, Jr., and K.W. Cummins. 1986. Ecology of coarse woody debris in temperate ecosystems. *Advances in Ecological Research* 15:133-302.

Hartel, P.G., J.D. Sumner, J.L. Hill, J.V. Collins, J.A. Entry, and W.I. Segars. 2002. Geographic variability of Escherichia coli ribotypes from animals in Idaho and Georgia. *Journal of Environmental Quality* 31:1,273-1,278.

Heimlich, R.E. 1994. Costs of an agricultural wetland reserve. *Land Economics* 70:234-246.

Heimlich, R. 2003. Agricultural resources and environmental indices. Agricultural Handbook No. 722. U.S. Department of Agriculture, Washington., D.C. [www.ers.usda.gov/publications/arei/ah722 accessed 8-21-06].

Hewlett, J.D., and A.R. Hibbert. 1967. Factors affecting the response of small watersheds to precipitation in humid regions. *Forest Hydrology* 275-290.

Hewlett, J.D., and J.C. Fortson. 1982. Stream temperature under an inadequate buffer strip in the southeast piedmont. *Water Resources Bulletin* 18:983-988.

Hively W.D., P. Gerard-Marchant, and T.S. Steenhuis. 2006. Distributed hydrological modeling of total dissolved phosphorus transport in an agricultural landscape, part II: dissolved phosphorus transport. *Hydrology and Earth Systems Sciences* 10(2): 263-276.

House. F. 2000. Totem salmon: Life lessons from another species. Beacon Press. Boston, Massachusetts. p. 248.

Jedrych, A.T., B.M. Olson, S.C. Nolan, and J.L. Little. 2006. Calculation of soil phosphorus limits for agricultural land in Alberta. In Alberta Soil Phosphorus Limits Project. Volume 2: Field-scale Losses and Soil Limits.

Alberta Agriculture, Food and Rural Development, Leth-bridge, Alberta. 87 pp.

Johansson, R.C., and J. Randall. 2003. Watershed abatement costs for agricultural phosphorus. *Water Resources Research* 39(4):1,088.

Jongman, R., and G. Pungetti, editors. 2004. Ecological networks and greenways: Concept, design, implementation. Cambridge University Press, Cambridge, England.

Kareiva, P., and M. Marvier. 2003. Conserving biodiversity coldspots. *American Scientist* 91: 344-348.

Karr, J.R., and I.J. Schlosser. 1978. Water resources at the land-water interface. *Science* 201:229-234.

Khanna, M., W. Yang, R. Farnsworth, and H. Onal. 2003. Optimal targeting of CREP to improve water quality: Determining land rental offers with endogenous sediment deposition coefficients. *American Journal of Agricultural Economics* 85(3): 538-553.

Khanna, M. and R.L. Farnsworth. 2005. Economics analysis of green payment policies for water quality. In R. Goetz and D. Berga, editors, Frontiers in Water Resource Economics. Kluwer Academic Publishers.

Kaspar, T.C., T.S. Colvin, D.B. Jaynes, D.E. James, D.W. Meek, D. Pulido, and H. Butler. 2003. Relationship between six years of corn yields and terrain attributes. *Precision Agriculture* 4:87-101.

Kendall, C. 1998. Tracing nitrogen sources and cycling in catchments. In C. Kendall and J.J. McDonnell, editors, Isotope Tracers in Catchment Hydrology, Elsevier, Amsterdam, The Netherlands. pp. 519-576.

Kimoto, A., M.A. Nearing, X.C. Zhang, and D.M. Powell. 2006. Applicability of rare earth element oxides as a sediment tracer for coarse-textured soils. *Catena* 65(3): 214-221.

Kravchenko, A.N., and D.G. Bullock. 2000. Correlation of corn and soybean grain yield with topography and soil properties. *Agronomy Journal* 92:75-83.

Kung, K-J.S. 1990. Preferential flow in sandy vadose zone: 2. Mechanism and implications. *Geoderma* 46:59-71.

Kuo, W-L., T.S. Steenhuis, C.E. McCulloch, C.L. Mohler, D.A. Weinstein, S.D. DeGloria, and D.P. Swaney. 1999. Effect of grid size on runoff and soil moisture for a variable-source-area hydrology model. *Water Resources Research* 35(11):3,419-3,428.

Little, J.L., S.C. Nolan, and J.P. Casson. 2006. Relationships between soil-test phosphorus and runoff phosphorus in small Alberta watersheds. In Alberta Soil Phosphorus Limits Project. Volume 2: Field-scale Losses and Soil Limits. Alberta Agriculture, Food and Rural Development, Lethbridge, Alberta. 150 pp.

Loftis, J.C., L.H. MacDonald, S. Streett, H.K. Iyer, and K. Bunte. 2001. Detecting cumulative watershed effects: The statistical power of pairing. *Journal of Hydrology* 251:49-64.

Lovegreen M. and others. 2006. personal contact. Bradford County Conservation District, Towanda, Pennsylvania., cite visits to Bentley and Mill Creeks August 1, 2006.

Lovejoy, T.E., and L. Hannah, editors. 2005. Climate change and biodiversity. Yale University Press, New Haven, Connecticutt.

Lowrance, R. L.S. Altier, J.D. Newbold, R.R. Schnabel, P.M. Groffman, J.M. Denver, D.L. Correll, J.W. Gilliam, J.L. Robinson, R.B. Brinsfield, K.W. Staver, W. Lucas, and A.L. Todd. 1997. Water quality functions of riparian forest buffers in Chesapeake Bay watersheds. *Environmental Management* 21:687-712.

Liu, P.L., J.L. Tian, P.H. Zhou, M.Y. Yang, and H. Shi. 2004. Stable rare earth element tracers to evaluate soil erosion. *Soil & Tillage Research* 76(2): 147-155.

Lyon, S.W., P. Gérard-Marcant, M.T. Walter, and T.S. Steenhuis. 2004. Using a topographic index to distribute variable source area runoff predicted with the SCS-Curve Number equation. *Hydrological Processes* 18(15): 2757-2771.

Lyon, S.W., J. Seibert, A.J. Lembo, M.T. Walter, and T.S. Steenhuis. 2006a. Geostatistical investigation into the temporal evolution of spatial structure in a shallow water table. *Hydrology and Earth System Sciences* 10: 113-125.

Lyon, S.W., A.J. Lembo, M.T. Walter, T.S. Steenhuis. 2006b. Linking science and application through user-friendly, internet-based GIS tools: Just Google it! *EOS* 87(38): 386.

Lyon, S.W., A.J. Lembo, M.T. Walter, and T.S. Steenhuis. 2006c. Defining probability of saturation with indicator kriging on hard and soft data. *Advances in Water Resources* 29(2): 181-193.

Maas, R.P., M.D. Smolen, and S.A. Dressing. 1985. Selecting critical areas for nonpoint-source pollution control. *Journal of Soil and Water Conservation* 40(1):68-71.

Mahler B.J., M. Winkler, P. Bennett, and D.M. Hillis. 1998. DNA-labeled clay: A sensitive new method for tracing particle transport. *Geology* 26(9): 831-834.

Mankin, K.R., P. Tuppard, D.L. Devlin, K.A. McVay, and W.L. Hargrove. 2005. Strategic targeting of watershed management using water quality modeling. In C.A. Brebbia and J.S. Antunes, editors, River Basin Management III. Transactions of Ecology and the Environment, Volume 83. WIT Press, Southampton, United Kingdom. Pp. 327-338

Margules, C. 2005. Conservation planning at the landscape scale. In J. Wiens and M. Moss, editors, Issues and Perspectives in Landscape Ecology. Cambridge University Press, Cambridge, England. pp. 230-237.

Margules, C., and R.L. Pressey. 2000. Systematic conservation planning. *Nature* 405: 243-253.

McClain, M.E., E.W. Boyer, C.L. Dent, S.E. Gergel, N.B. Grimm, P.M. Groffman, S.C. Hart, J.W. Harvey, C.A. Johnston, E. Mayorga, W.H. McDowell, and G. Pinay.

2003. Biogeochemical hot spots and hot moments at the interface of terrestrial and aquatic ecosystems. *Ecosystems* 6: 301-312.

McDonnell, J.J. 2005. Discussion of "simple estimation of prevalence of Hortonian flow in New York City watersheds." By M.T Walter, V.K. Mehta, A.M. Marrone, J. Boll, P. Gerard-Marchant, T.S. Steenhuis, and M.F. Walter. *American Society of Civil Engineers, Journal of Hydrologic Engineering* 10(2): 168-169.

McGlynn, B.L., and J.J. McDonnell. 2003. Quantifying the relative contributions of riparian and hillslope zones to catchment runoff. *Water Resources Research* 39(11).

McGarigal, K., S.A. Cushman, M.C. Neel, and E. Ene. 2002. FRAGSTATS: *Spatial pattern analysis for categorical maps.* University of Massachusetts, Amherst.

Meals, D.W. 2001. Water quality response to riparian restoration in an agricultural watershed in Vermont, USA. *Water Science and Technology.* 43(5): 175-182.

Mehta, V.K., M.T. Walter, E.S. Brooks, T.S. Steenhuis, M.F. Walter, M. Johnson, J. Boll, and D. Thongs. 2004. Evaluation and application of SMR for watershed modeling in the Catskill Mountains of New York State. *Environmental Modeling & Assessment* 9(2): 77-89.

Millennium Ecosystem Assessment. 2005. Ecosystems and human well-being: Synthesis. Island Press, Washington D.C.

Miller, J.R., and R.J. Hobbs. 2002. Conservation where people live and work. *Conservation Biology* 16: 330-337.

Mittermeier, R.A., N. Myers, J.G. Thomsen, G.A. da Fonseca, and S. Oliveri. 1998. Biodiversity hotspots and major tropical wilderness areas: approaches to setting conservation priorities. *Conservation Biology* 12: 516-520.

Moore, I.D., R.B. Grayson, and A.R. Ladson. 1991. Digital terrain modeling: A review of hydrological, geomorphological, and biological applications. *Hydrological Processes* 5:3-30.

Morrison, M.L., and L.S. Hall. 2002. Standard terminology: Toward a common language to advance ecological understanding and application. In J.M. Scott, P.J. Heglund, M.L. Morrison, J.B. Haufler, M.G. Raphael, W.A. Wall, and F.B. Sampson, editors, Predicting Species Occurrences: Issues of Accuracy and Scale.Island Press, Washington, D.C. pp. 43-52.

Myers, N., R. Mittermeier, C.G. Mittermeier, G.A.B. da Fonseca, and J. Kent. 2000. Biodiversity hotspots for conservation priorities. *Nature* 403: 853-858.

Nagle G.N., and J.C. Ritchie. 1999. The use of tracers to study sediment sources in three streams in northeastern Oregon. *Physical Geography* 20(4): 348-366.

Nagle G.N., and J.C. Ritchie. 2004. Wheat field erosion rates and channel bottom sediment sources in an intensively cropped northeastern Oregon drainage basin. *Land Degradation & Development* 15(1): 15-26.

Nagle, G.N., T.J. Fahey, J.C. Ritchie, and P.B. Woodbury. 2006. Variations in sediment sources and yields in the Finger Lakes and Catskills Regions of New York. *Hydrological Processes* (in press)

Naiman, R.J., H. Decamps, and M. Pollock. 1993. The role of riparian corridors in maintaining regional biodiversity. *Ecological Applications* 3:209-212.

National Research Council. 1993. Soil and water quality: An agenda for agriculture. National Academy Press, Washington, D.C.

National Research Council. 2002. Riparian areas: Functions and strategies for management. National Academy Press, Washington, D.C. 428 pp.

Newbold, J.D., D.C. Erman, and K.B. Roby. 1980. Effects of logging on macroinvertebrates in streams with and without buffer strips. *Canadian Journal of Fisheries and Aquatic Sciences* 37:1,077-1,085.

Noguchi S., Y. Tsuboyama, R.C. Sidle, and I. Hosoda. 2001. Subsurface runoff characteristics from a forest hillslope soil profile including macropores, Hitachi Ohta, Japan. *Hydrological Processes* 15(11): 2131-2149.

New York City Department of Environmental Protection. 1991. Ad Hoc Task Force on Agricultural and New York City Watershed Regulations, policy group recommendations. New York City Department of Environmental Protection, Elmhurst. 16 pp + appendix.

New York State Water Resources Insitute. 1992. Watershed protection program for the Catskill-Delaware-Croton system, New York City. New York State Water Resources Institute, Cornell Univeristy, Ithaca. 96 pp.

O'Loughlin, E.M. 1981. Saturation regions is catchments and their relations to soil and topographic attributes. *Journal of Hydrology* 53:229-246.

O'Loughlin, E.M. 1986. Prediciton of surface saturation zones in natural catchments by topographic analysis. *Water Resources Research* 22(5):794-804.

Olson, D.M. and E. Dinerstein. 1998. The Global 200: a representation approach to conserving the earth's most biologically valuable ecoregions. *Conservation Biology* 12: 502-515.

Peterson, D.L. and V.T. Parker, editors. 1998. *Ecological scale: Theory and applications.* Columbia University Press, New York, New York.

Polasky, S., J.D. Camm, and B. Garber-Yonts. 2001. Selecting biological reserves cost-effectively: An application to terrestrial vertebrate conservation in Oregon. *Land Economics* 77(1): 68-78.

Qiu, Z.Y. 2003. A VSA-based strategy for placing conservation buffers in agricultural watersheds. *Environmental Management* 32 (3): 299-311.

Rallison, R.K. 1980. Origin and evolution of the SCS runoff equation. In Proceedings of Symposium on Watershed

Management, 21–23 July, Boise, Idaho. American Society of Civil Engineers, New York, New York. Pp. 912–924.

Randall, G.W., and D.J. Mulla. 2001. Nitrate nitrogen in surface waters as influenced by climatic conditions and agricultural practices. *Journal of Environmental Quality* 30(2):337-344.

Reiners, W.A., and K.L. Driese. 2004. Transport processes in nature: Propagation of ecological influences through environmental space. Cambridge University Press, Cambridge, England.

Rejesus, R.M., and R.H. Hornbaker. 1999. Economic and environmental evaluation of alternative pollution-reducing nitrogen management practices in central Illinois. *Agriculture, Ecosystems & Environment* 75(1):41-53.

Rosenblatt, A.E., A. J. Gold, M.H. Stolt, P.M. Groffman, and D.Q. Kellogg. 2001. Identifying riparian sinks for watershed nitrate using soil surveys. *Journal of Environmental Quality* 30:1596-1604.

Russell, M.A., D.E. Walling, and R.A. Hodgkinson. 2001. Suspended sediment sources in two small lowland agricultural catchments in the UK. *Journal of Hydrology* 252:1-24.

Schneiderman, E.M., T.S. Steenhuis, D.J. Thongs, Z.M. Easton, M.S. Zion, A.L. Neal, G.F. Mendoza, and M.T. Walter. 2006. Incorporating variable source area hydrology into Curve Number based watershed loading functions. *Hydrological Processes* (in press).

Schultz, R.C., J.P. Colletti, T.M. Isenhart, W.W. Simpkins, C.W. Mize, and M.L. Thompson. 1995. Design and placement of a multi-species riparian buffer strip system. *Agroforestry Systems* 29:201-226.

Scott, J.M., P.J. Heglund, M.L. Morrison, J.B. Haufler, M.G. Raphael, W.A. Wall, and F.B. Sampson, editors. 2002. Predicting species occurrences: Issues of accuracy and scale. Island Press, Washington D.C.

Sharpley, A. J.J. Meisinger, A. Breeuwsma, J.T. Sims, T.C. Daniel, and J.S. Schepers. 1998. Impacts of animal manure management on ground and surface water quality. In J.L. Hatfield and B.A. Stewart, editors, Animal Waste Utilization: Effective Use of Manure as a Soil Resource. Ann Arbor Press, Chelsea, Michigan. Pp. 173-242.

Schlosser, I.J. 1991. Stream fish ecology: a landscape perspective. *BioScience* 41:704-712.

Sidle R.C., S. Noguchi, Y. Tsuboyama , and K. Laursen. 2001. A conceptual model of preferential flow systems in forested hillslopes: evidence of self-organization. *Hydrological Processes* 15(10): 1,675-1,692.

Simpkins, W.W., T.R. Wineland, R.J. Andress, D.A. Johnston, G.C. Caron, T.M. Isenhart, and R.C. Schultz. 2002. Hydrogeological constraints on riparian buffers for reduction of diffuse pollution:examples from the Bear

Creek watershed in Iowa, USA. *Water Science and Technology* 45(9):61-68.

Souchere, V., D. King, J. Daroussin, F. Papy, and A. Capillon. 1998. Effects of tillage on runoff directions: consequences on runoff contributing area within agricultural catchments. *Journal of Hydrology* 206:256-267.

Stauffer, D.F. 2002. Linking populations and habitats: Where have we been? Where are we going? In J.M. Scott, P.J. Heglund, M.L. Morrison, J.B. Haufler, M.G. Raphael, W.A. Wall, and F.B. Sampson, editors, Predicting Species Occurrences: Issues of Accuracy and Scale. Island Press, Washington, D.C. pp. 53-62.

Steenhuis, T.S., M. Winchell, J. Rossing, J.A. Zollweg, and M.F. Walter. 1995. SCS runoff equation revisited for variable-source runoff areas. *American Society of Civil Engineers, Journal of Irrigation and Drainage Engineering* 121: 234-238.

Soulsby, C., J. Petry, M.J. Brewer, S.M. Dunn, B. Ott, and I.A. Malcolm. 2003. Identifying and assessing uncertainty in hydrological pathways: a novel approach to end member mixing in a Scottish agricultural catchment. *Journal of Hydrology* 274:109-128.

Theobald, D.M. 2005. Landscape patterns of exurban growth in the USA from 1980 to 2020. *Ecology and Society* 10(1): 32. http://www.ecologyandsociety.org/vol10/iss1/art32/

Thompson, J.A., J.C. Bell, and C.A Butler. 1997. Quantitative soil landscape modeling for estimating the areal extent of hydromorphic soils. *Soil Science Society of America Journal* 61:971-980.

Tomer, M.D., D.E. James, and T.M. Isenhart. 2003. Optimizing the placement of riparian practices in a watershed using terrain analysis. *Journal of Soil and Water Conservation* 58(4):198-206.

Tomer, M.D., and D.E. James. 2004. Do soil surveys and terrain analyses identify similar priority sites for conservation? *Soil Science Society of America Journal* 68(6):1905-1915.

Tomer, M.D., M.G. Dosskey, M.R. Burkart, D.E. James, M.J. Helmers, and D.E. Eisenhauer. 2006. Methods to prioritize placement of riparian buffers for improved water quality. *Agroforestry Systems* (in press).

Trimble, S.W. 1999a. Dcreased rates of alluvial sediment storage in the Coon Creek Basin, Wisconsin, 1975-993. *Science* 285:1,244-1,246.

Trimble, S.W. 1999b. Response to comment on "Decreased rates of alluvial sediment storage in the Coon Creek Basin, Wisconsin, 1975-1993" by Pimentel & Skidmore. *Science* 286: 1,477-1,478

Trimble, S.W., and P. Crosson. 2000a. U.S. soil erosion rates--myth and reality. *Science* 289:248-250.

Trimble, S.W., and P. Crosson. 2000b. Response to comment on U.S. soil erosion rates--myth and reality by Nearing et al. *Science* 290: 1,301.

Trimble, S.W. 2004. Effects of riparian vegetation on stream channel stability and sediment budgets. In S.J. Bennett and A. Simon, editors, Riparian Vegetation and Fluvial Morphology. American Geophysical Union, Washington D.C. pp. 153-170

Turner, M.G., R.H. Gardner, and R.V. O'Neill. 1995. Ecological dynamics at broad scales. *BioScience Supplement* S-29 to S-35.

Uhlenbrook, S., and S. Hoag. 2003. Quantifying uncertainties in tracer-based hydrograph separations: a case study for two-, three-, and five-component hydrograph separations in a mountainous catchment. *Hydrological Processes* 17:431-453.

U.S. Department of Agriculture-Farm Service Agency. 1999. Environmental benefits index. Fact sheet: Conservation Reserve Program sign-up 20. U.S. Department of Agriculture, Washington, D.C. 6 pp.

U.S. Department of Agriculture-Farm Service Agency. 2003. Conservation Reserve Program: Final programmatic environmental impact statement. January (http://www. fsa.usda.gov/-dafp/cepd/epb/impact.htm#final). Accessed November 17, 2003.

U.S. Department of Agriculture-Natural Resources Conservation Service. 1994. The phosphorus index: A phosphorus assessment tool. Technical Note. Series No. 1901. Web: <http://www.nhq.nrcs.usda.gov/ BCS/nutri/phosphor.html>

U.S. Department of Agriculture-Natural Resources Conservation Service. 1999. Conservation corridor planning at the landscape level: Managing for wildlife habitat. Part 614.4 National Biology Handbook, Part 190. Natural Resources Conservation Service, U.S. Department of Agriculture, Washington, D.C.

U.S. Department of Agriculture-Soil Conservation Service. 1972. National Engineering Handbook, Part 630 Hydrology, Section 4, Chapter 7.

Vannote, R.L., G.W. Minshall, K.W. Cummins, J.R. Sedell, and C.E. Cushing. 1980. The river continuum concept. *Canadian Journal of Fisheries and Aquatic Sciences* 37:130-137.

Veith, T. L., M. L. Wolfe, and C. D. Heatwole. 2003. Development of optimization procedure for cost-effective BMP placement. J. *American Water Resources Association* 39(6): 1,331-1,343.

Veith, T.L., M.L. Wolfe, and C.D. Heatwole. 2004. Cost-effective BMP placement: Optimization versus targeting. Transactions, *American Society of Agricultural Engineers* 47(5): 1,585-1,594.

Verhoest, N.E.C., P.A. Troch, C. Paniconi, and F.P De Troch. 1998. Mapping basin scale variable source areas from multitemporal remotely sensed observations of soil moisture behavior. *Water Resources Research* 34(12):3,235-3,244.

Vertessy, R.A., and H. Elsenbeer. 1999. Distributed modeling of storm flow generation in an Amazonian rain forest catchment: Effects of model parameterization. *Water Resources Research* 35(7):2,173-2,187.

Vidon, P., and A.R. Hill. 2006. A landscape based approach to estimate riparian hydrological and nitrate removal functions. *Journal of the American Water Resources Association* 42(4):1,099-1,112.

Wallbrink, P.J., A.S. Murray, and J.M. Olley. 1999. Relating suspended sediment to its original soil depth using fallout radionuclides. *Soil Science Society of America Journal* 63(2):369-378.

Walter, M.F., T.S. Steenhuis, and D.A. Haith. 1979. Nonpoint source pollution control by soil and water conservation practices. *Transactions, American Society of Agricultural Engineers* 22(5): 834-840.

Walter, M.T., and M.F. Walter. 1999. The New York City Watershed Agricultural Program (WAP): A model for comprehensive planning for water quality and agricultural economic Viability. *Water Resources Impact* 1(5): 5-8.

Walter, M.T., M.F. Walter, E.S. Brooks, T.S. Steenhuis, J. Boll, and K.R. Weiler. 2000. Hydrologically sensitive areas: Variable source area hydrology implications for water quality risk assessment. *Journal of Soil and Water Conservation* 3: 277-284.

Walter, M.T., E.S. Brooks, M.F. Walter, T.S. Steenhuis, C.A. Scott, and J.Boll. 2001. Evaluation of soluble phosphorus transport from manure-applied fields under various spreading strategies. *Journal of Soil & Water Conservation* 56(4): 329-336.

Walter, M.T., V.K. Mehta, A.M. Marrone, J. Boll, P. Gérard-Merchant, T.S. Steenhuis, and M.F. Walter. 2003. A simple estimation of the prevalence of Hortonian flow in New York City's watersheds. *American Society of Civil Engineers, Journal of Hydrologic Engineering* 8(4): 214-218.

Walter, M.T., P. Gérard-Merchant, T.S. Steenhuis, and M.F. Walter. 2005. Closure: A simple estimation of the prevalence of Hortonian flow in New York City's watersheds. *American Society of Civil Engineers, Journal of Hydrologic Engineering* 10(2): 169-170.

Walter, M.T., and S.B. Shaw. 2005. Discussion: "Curve number hydrology in water quality modeling: Uses, abuses, and future directions" by Garen and Moore. *Journal of the American Water Resources Association* 41(6): 1,491-1,492.

Wang, L.Z., J. Lyons, and P. Kanehl. 2002. Effects of watershed best management practices on habitat and fish in Wisconsin streams. *Journal of the American Water Resources Association* 38(3): 663-680.

Welsh H.H., G.R. Hodgson, and A.J. Lind. 2005. Eco-geography of the herpetofauna of a northern California watershed: linking species patterns to landscape processes. *Ecography* 28(4): 521-536.

Western A.W., S.L. Zhou, R.B. Grayson, T.A. McMahon, G. Bloschl, and D.J. Wilson. 2004. Spatial correlation of soil moisture in small catchments and its relationship to dominant spatial hydrological processes. *Journal of Hydrology.* 286(1-4): 113-134.

Wiens, J.A. 1989. Spatial scaling in ecology. *Functional Ecology* 3: 385-397.

Wiens, J.A. 2001. The landscape context of dispersal. In J. Clobert, E. Danchin, A.A. Dhondt. and J.D. Nichols, editors, Dispersal. Oxford University Press, Oxford, England. pp. 96-109.

Wiens, J.A., B. Van Horne, B.R. Noon. 2002. Integrating landscape structure and scale into natural resource management. In J. Liu and W.W. Taylor, editors, Integrating Landscape Ecology into Natural Resource Management. Cambridge University Press, Cambridge, England. pp. 23-67.

Wiggins, B.A., P.W. Cash, W.S. Creamer, S.E. Dart, P.G. Garcia, T.M Gerecke, J. Han, B.L. Henry, K.B. Hoover, E.L. Johnson, K.C. Jones J.G. McCarthy, J.A. McDonough, S.A. Mercer, M.J. Noto, H. Park, M.S. Phillips, S.M. Purner, B.M. Smith, E.N. Stevens, and A.K. Varne. 2004. Use of antibiotic resistance analysis for representativeness testing of multiwatershed libraries. *Applied and Enviromental Microbiology* 69(6): 3,399-3,405.

Wilcove, D.S., D. Rothstein, J. Dubow, A. Phillips, and E. Losos. 1998. Quantifying threats to imperiled species in the United States. *BioScience* 48: 607-615.

Wilson, G.V., P.M. Jardine, R.J. Luxmoore, and J.R. Jones. 1990. Hydrology of a forested hillslope during storm events. *Geoderma* 46:119-138.

Wilson, K.A., M. McBride, M. Bode, and H.P. Possingham. 2006. Prioritising global conservation efforts. *Nature* 440: 337-340.

Wischmeier, W.H., and D.D. Smith. 1978. Predicting rainfall erosion losses – a guide to conservation planning. Agricultural Handbook No. 537. U.S. Department of Agriculture, Washington, D.C.

With, K.A., and A.W. King. 1997. The use and abuse of neutral landscape models in ecology. *Oikos* 79: 219-229.

Wu, J., and B.A. Babcock. 1999. Metamodeling potential nitrate water pollution in the central United States. *Journal of Environmental Quality* 28:1,916-1,928.

Yang W.H., and A. Weersink. 2004. Cost-effective targeting of riparian buffers. *Canadian Journal of Agricultural Economics-Revue Canadienne D Agroeconomie* 52(1): 17-34.

Young, R.A., C.A. Onstad, D.D. Boesch, and W.P. Anderson. 1989. AGNPS – a nonpoint-source pollution model for evaluating agricultural watersheds. *Journal of Soil and Water Conservation* 44(2): 168-173.

Roundtable:
The science of targeting to improve conservation effectiveness

Frustration exists among many natural resource professionals over the difficulty of moving forward with targeting conservation in watersheds and on landscapes. Successful targeting in watersheds and on landscapes depends upon both biophysical and social factors:

- Biophysical information that identifies sensitive areas within landscapes.

- Behavioral information that explains why inappropriate land management occurs in sensitive areas and identifies factors that determine the willingness of land managers to adopt conservation in those locations.

- Program structure that encourages conservation adoption by landowners and managers in targeted areas.

- Policy that motivates professionals to target conservation to owners and managers of sensitive areas where greater environmental impact can be achieved.

Biophysical models appear well advanced and continue to refine our ability to identify sensitive areas within landscapes and watersheds. Existing models are underutilized, and emerging approaches and technologies promise to enable even greater conservation efficiency. But the power of biophysical models will be widely applied only if the knowledge contained in complex, research-type models is translated into simplified models that are easily used by natural resource professionals. This translation must be accompanied by assessments of uncertainty and limitations in the models and guidance on selection and use of appropriate spatial data from a rapidly growing and complex body of available sources.

Greater challenges involve the human dimensions of targeting where difficulties stem from behaviors of both land managers and natural resource professionals. In many cases, great strides in conservation effectiveness could be made by changing the management behaviors of a small fraction of land managers within watersheds or landscapes. Motivating these few land managers to adopt conservation practices by way of education and financial incentives has not been very successful, which suggests that there are other important factors that determine the willingness or unwillingness of key land managers to adopt conservation practices. Better understanding of the broader suite of motivational factors would guide improvements in incentive strategies. Effective strategies must encourage land managers to choose practices that accrue conservation benefits mainly to society (or to the watershed) over those that provide benefits mainly to the individual farm or ranch.

Policy and programs need to be structured in ways that encourage targeted application of conservation practices at the appropriate scale. Today, targeting often amounts to random application of conservation within large-scale problem areas, such as watersheds with total maximum daily load (TMDL) concerns, whereas finer-scale targeting may be necessary to achieve meaningful conservation impacts. Important strategies remain underutilized, such as flexibility in tailoring enrollment criteria, like the environmental benefits index, to meet local needs, and marketing conservation practices to managers in targeted areas while maintaining equal-access requirements. Greater use of existing targeting strategies could be achieved if there were policies that rewarded targeting efforts among natural resource professionals.

Finally, greater local control of targeting decisions may further improve conservation success. Local knowledge of sensitive areas and management behaviors can provide critical information at finer scales of resolution than generalized biophysical and behavioral models.

Top-down decision-making based on generalized models and enrollment criteria may be too coarse to identify key locations and motivate those land managers who can produce the greatest conservation impact.

Buying environmental services: Effective use of economic tools

Roger Claassen

Environmental targeting in voluntary agricultural conservation programs involves identifying land and practices that can yield large environmental benefits relative to the payment used to induce practice adoption. When funding for conservation is limited, a program that obtains the largest possible benefit per dollar of program expenditure is fully "cost-effective" (see Cattaneo et al., 2005, for a discussion of various concepts of cost-effectiveness). The level of cost-effectiveness actually achieved, however, can vary greatly with the details of program design. Effectiveness in identifying potential environmental benefits is obviously important. Of equal importance is the level of payment used to obtain those benefits.

The level of payment is important in at least two respects. First, because conservation program budgets are limited and likely to continue to be limited, payments that exceed the level needed to secure a producer's participation effectively reduce the number of producers, acres, and practices that can be enrolled. Thus, the efficiency with which programs convert budget to environmental quality depends upon payment amounts. Second,

because most conservation programs are voluntary, participation depends upon sufficient incentives. How payment incentives are structured and how large they are can have a significant impact on which producers apply, what land they are willing to enroll, and which practices they are willing to adopt on that land.

The interaction of payment incentives and environmental benefit-cost targeting is critical to obtaining cost-effective environmental gains. U.S. Department of Agriculture (USDA) conservation programs generally specify a process for identifying the producers, land, and practices that will eventually be enrolled in the program. The juxtaposition of payment incentives and targeting mechanisms within this broader process is critical to identifying producers who can make a cost-effective contribution to improving environmental quality.

Program design as a winnowing process

Conservation program enrollment can be viewed as a winnowing process (Figure 1; Claassen et al., 2005). Beginning with all agricultural producers and land, various mechanisms are used to narrow the set of possible producers and land down to those eventually enrolled in the program and determine which practices will be applied and where. How much winnowing must be done depends largely upon the program constraint, either budget cap or acreage cap, depending upon the program. How the winnowing happens largely determines the level of environmental gain that will be achieved, given the budget or acreage cap. I refer to the various facets of this process collectively as "program design."

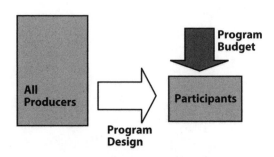

Figure 1. Program enrollment as a winnowing process

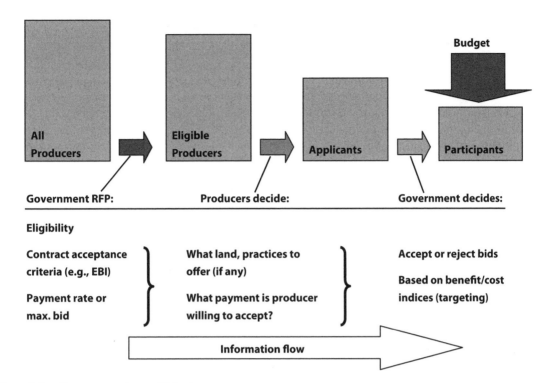

Figure 2. Enrollment process has multiple steps

In most conservation programs, the process begins with a request for proposals (RFP). The government's RFP, often referred to as a "sign-up notice," spells out the details of the government's request in much the same way governments spell out requirements when bidding out road projects or other public works (Cattaneo et al., 2005). The RFP provides information on (1) who can submit proposals (eligibility), (2) the type of land and practices the government would like to enroll, (3) how much farmers can expect to be paid (or, in some programs, the maximum bid that could be accepted), and (4) how producer applications will be assessed. Bid assessment often involves use of a benefit-cost index to rank producer applications.

The first step in the winnowing process is to determine which producers, land, and practices are eligible (Figure 2). Eligibility criteria vary across programs. In the Environmental Quality Incentives Program (EQIP), for example, eligibility is broad. Most farmers may apply for EQIP enrollment, even though not all farms have the same chance of being selected for participation. Moreover, EQIP applicants can offer any type of agricultural land and any of a wide range of practices so long as the practices address a recognized resource concern (environmental problem) on the farm. In other programs, eligibility is used as a first (albeit crude) targeting mechanism. In general signups for the Conservation Reserve Program (CRP), for example, eligibility is limited to cropland that is highly erodible, located in a conservation priority area, or land under an expiring CRP contract (U.S. Department of Agriculture-Farm Service Agency, 2006).

The second step in the winnowing process is producer application. Here, eligible producers (not the government) decide which land, if any, to offer for program enrollment, what practices they are willing to adopt or install on that land, and what payment they are willing to accept for taking these actions. Producers often have broad latitude regarding the land and practices they propose in contract offers. For example, a producer might offer to address a local water quality concern by reducing soil erosion or nutrient runoff from fields or by installing a field-edge filter strip or grassed waterway to capture sediment and nutrients before they leave the farm. Alternately (or in addition), the producer may offer to

Table 1. Benefit-cost targeting in the CRP: The Environmental Benefits Index (EBI) (general signup 26)

EBI Factors	Definition	Features that Increase Points	Maximum Points
Wildlife	Evaluates the expected wildlife benefits of the offer	• Diversity of grass/legumes • Use of native grasses • Tree planting • Wetlands restoration • Beneficial to threatened/endangered species • Complements wetland habitat	100
Water Quality	Evaluates the potential surface and ground water impacts	• Located in ground- or surface-water protection area • Potential for percolation of chemicals and the local population using groundwater • Potential for runoff to reach surface water and the county population	100
Erosion	Evaluates soil erodibility	• Larger field-average erodibility index	100
Enduring Benefits	Evaluates the likelihood for practice to remain	• Tree cover • Wetland restoration	50
Air Quality	Evaluates gains from reduced dust	• Potential for dust to affect people • Soil vulnerability to wind erosion • Carbon sequestrations	45
Cost	Evaluates cost of parcel	• Lower CRP rent • No government cost share • Payment is below program's maximum acceptable for area and soil type	Varies

Note: This table includes the most common and highest scoring practices. For more information, see USDA, FSA, 2004.

develop wildlife food plots or otherwise alter production practices to enhance wildlife habitat.

Producer application decisions are likely to be informed by a description of ranking criteria and potential payments provided in the RFP. One example of a ranking mechanism is the environmental benefits index (EBI) used in the CRP general signup. While the EBI has changed some over the years, it primarily focuses on enhancing wildlife habitat, improving water quality, and maintaining soil productivity (Table 1). Producers receive detailed information on the EBI, including the relative weight given to addressing various resource concerns, for example, wildlife, water quality, and soil quality. The RFP also provides information on the level of payment producers are likely to receive. Depending upon the program, producers may be told what the payment

rate is, what level of (percentage) cost-sharing they can receive, or, if producers are asked to submit a bid for the level of financial assistance they would accept, the maximum acceptable bid. For example, producers applying for CRP enrollment (through a general sign-up) are asked to bid on the level of annual payments (dollars per acre) they are willing to accept, up to the bid limit established for the field being offered. Applicants who offer a bid below the maximum acceptable bid improve their chance of being selected for enrollment.

Which fields and what practices are offered will depend upon how each would be treated in the program's ranking mechanism (e.g., environmental benefit-cost index) and the level of payment being offered for each specific action. More specifically, the producer may consider (1) how the specific details of the offer (including any bid for financial assistance) will affect his chances of program enrollment and (2) whether the payment being offered (or the maximum bid) is sufficient to cover out-of-pocket costs, lost production, increased risk, etc. The producer will make an offer only if he or she is willing to accept the level of payment being offered by the government or, if bids on financial assistance are being solicited, the maximum acceptable bid is greater than what the producer is willing to accept.

In the final step, the government decides which bids to accept and which to reject, often using a benefit-cost index, like the EBI, to rank contract offers. Information flowing from applicants to USDA is critical in ranking applications for acceptance. Scoring the factors in the EBI and other benefit-cost indices requires field-specific information on soils, topography, and location, as well as information on the type of land cover that is proposed (e.g., grass versus trees), and the cost of the contract. Some of this information can be obtained directly from contract offers (e.g., proposed practices and cost), while other information can be added using the field location and existing geographic information system (GIS) databases (e.g., soils and topography).

Once contract offers are ranked, program managers must select a cut-off score. Contracts with scores above the cut off are accepted; contracts with scores below the cut-off are rejected. A critical consideration in setting the cut-off is the budget available to fund contracts. Many USDA conservation programs (e.g., EQIP) have fixed annual budgets that cannot be exceeded; this strictly limits the number of producer applications that can be accepted. Some programs, particularly the CRP, have overall acreage limitations that cannot be breached at any time during the life of the program. In the CRP, for example, the current limit is 39.2 million acres. Because the program operates on an overall limit, rather than annual limit, CRP program managers have not always accepted every contract offer that could have been funded under the acreage cap. Some contract offers that would yield only modest environmental gains in relation to contract costs, have been rejected because of the possibility that better contracts (more environmental benefits per dollar of cost) will be offered in subsequent sign-ups.

For many USDA programs, including CRP

and EQIP, these rankings are critical to environmental benefit-cost targeting. In these programs, the ranking is the only point at which the field-specific potential for environmental benefits is weighed against field-specific costs. As such, successful benefit-cost targeting depends directly upon the success of this process. Of course, design of the index itself is critical to the success of targeting: Which environmental benefits are emphasized, how point totals are derived, and how cost is incorporated into the index are all important issues.

A second critical issue is the composition of the applicant pool. The effectiveness of any ranking mechanism in selecting producers who can provide large environmental gains per dollar of conservation expenditure depends entirely upon the pool of applications that program managers have to choose from. In other words, producers who can offer cost-effective conservation can be selected for program participation only if they submit an application. The relationship between payment incentives and a producer's willingness to accept is critical: If the offered payment is less than the minimum payment the producer is willing to accept, there is little chance the producer will apply. While insufficient payments are not the only barrier to program application (information about available programs and/or environmental problems is also important), it can be a critical barrier. Barriers to program application, depending upon who they exclude, can result in an applicant pool that limits the environmental cost-effectiveness of the program.

A digression on willingness to accept

A producer's "willingness to accept" is the minimum payment he or she would accept in exchange for taking a specific action, such as adopting a specific conservation practice or groups of practices. Willingness to accept depends upon a range of factors, including out-of-pocket costs (or savings), changes in production (positive or negative), changes in production risk, and the level of management skill required for successful implementation. In other words, the level of compensation producers require for adoption of a given conservation practice depends upon the level of additional cost they incur, production losses they expect to incur, increased pro-

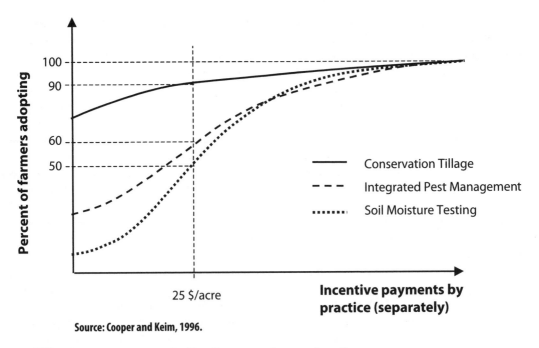

Source: Cooper and Keim, 1996.

Figure 3. Willingness to accept can vary significantly across producers and practices

duction risk (i.e., increased probability of production losses), and increased management effort.

Willingness to accept can vary widely among producers and practices. Cooper and Keim (1996) estimated the distribution of willingness to accept for a number of commonly applied management practices. In figure 3, the percentage of producers adopting conservation tillage, integrated pest management (IPM), or soil moisture testing (where appropriate) is a function of the per-acre payment offered (those with zero willingness to accept have already adopted the practice). A majority of producers surveyed for this study had already adopted conservation tillage, while smaller proportions had adopted IPM and soil moisture testing. Offered $10 per hectare ($25 per acre) for adoption of any of these practices, adoption rates would increase to 90, 60, and 50 percent for conservation tillage, IPM, and soil moisture testing, respectively. While this study did not involve a nationally representative sample, it does illustrate how willingness to accept can vary by practice and producer.

Producer willingness to accept is private information – it cannot be directly observed by the government (or anyone else). In some cases, program managers can gain insight by assessing the factors affecting willingness to accept, such as out-of-pocket costs or production changes. The extent to which these factors can be assessed, however, varies considerably across practices and producers. For many structural practices, such as terraces, grassed waterways, and filter strips, assessments may be relatively easy. When building terraces, for example, out-of-pocket costs largely involve the cost of surveying, dirt work, and installation of a system to drain terrace channels. Production and risk effects would be quite small, at least in the short run. Even so, willingness to accept will also depend upon the level of private benefit perceived by the producer, such as soil productivity maintenance through erosion control.

In other cases, such as the adoption of management practices, changes in cost and production are more difficult to assess. For management practices, changes in out-of-pocket costs are more difficult to assess because they may involve numerous changes in input purchase and machinery investment decisions that are difficult to separate from year-to-year variation in input and machinery expenses. Conservation tillage, for example, increases some costs while reducing others (Sandretto, 1997). Producers who switch to conservation tillage may save on labor, fuel, and net capital investment in equipment, but see increased cost for herbicides or fertilizer. Likewise, nutrient management may involve reduced fertilizer use, but increase the cost of application if additional equipment is needed or multiple applications are used to improve application timing. In both cases, the exact balance of additional cost and cost savings may vary across regions and farms.

Production and risk effects are even more difficult to attribute to practice changes because they can be difficult to separate from production fluctuations due to weather and spatial variations due to soil productivity, management skill, and other factors. For example, producers who apply more fertilizer than necessary to achieve yield goals may do so in part because they are uncertain about the level of fertilizer needed to achieve the yield goal (see Sheriff, 2005, for a survey of economic research on fertilizer application issues). The economically optimal level of fertilizer application (the level where returns to fertilizer application are maximized) may vary from year to year as fertilizer prices, crop prices, and growing conditions vary. Because crop prices and growing conditions are not known until the end of the growing season, producers always make fertilizer purchase decisions under uncertainty. Likewise, split application exposes producers to the risk that weather conditions will delay application at a time of high nutrient uptake by crops. At best, producers could understand the distribution of possible yield effects from reduced fertilizer application or split application and the probability that each will occur.

Finally, differences in education, experience, primary occupation (farm or nonfarm), and other socioeconomic characteristics can be important in the adoption of various conservation practices (e.g., Lambert et al., 2006; Caswell et al., 2001; Cooper and Keim, 1996; Soule et al., 2000; Khanna, 2001; Lichtenberg, 2004). For example, Lambert et al. argue that younger, better educated farmers, with larger farms, whose principal occupation is farming, are more likely to adopt practices that require a higher level of management skill or learning new skills, such as nutrient management. This higher likelihood of adoption may reflect a lower willingness to accept for these producers (even zero willingness to accept for those who have adopted without incentive payments).

Payment incentives and targeting: How are they related?

As already emphasized, payment incentives are critical in forming the applicant pool. Ranking mechanisms for benefit-cost targeting, in turn, can only be effective if the applicant pool includes the producers who can offer the combinations of land and practices that yield the largest environmental gain per dollar of expenditure.

A simple example, involving a hypothetical program, illustrates how payment mechanisms can affect program application and, therefore, benefit-cost targeting and the overall effectiveness of the conservation programs. Consider a hypothetical, single-objective program (e.g., reducing sediment loads to water) offered in a small area, such as a single watershed. Producers within the watershed would all take similar actions to address the program objective, but their willingness to accept payments and the environmental effectiveness of their actions will vary. In figure 4, each dot represents a single farm's potential for environmental gain (environmental score) and the producer's willingness to accept for taking prescribed conservation actions. The government decides what payment to offer and how to rank applications for participation, assuming that some must be rejected because of budget constraints. While this example is unrealistically simple, it allows focus on basic concepts by stripping away the complexities found in real conservation programs. More general methods, similar to those described here, have been used in actual conservation programs.

If the government offers a single, uniform payment (on a per-acre or other basis, as appropriate) for adopting or installing the specified practice(s), the applicant pool is comprised of producers with a willingness to accept less than the payment rate (farms in the shaded area in figure 5). Some producers who have willingness to accept well below the payment rate may find participation quite profitable. A benefit-cost index, similar (at least in spirit) to the EBI, could be used to determine which offers are accepted. The "index cutoff" line in figure 6 is a locus of points with a constant index score. The line slopes upward to the right because higher payments are appropriate in exchange for greater environmental gain (as measured by the environmental score). Applicants with scores above a cutoff level would be

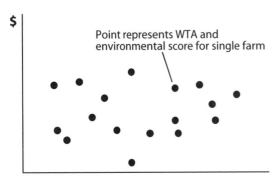

Figure 4. Willingness to accept (WTA) and environmental gain for farms in a small watershed

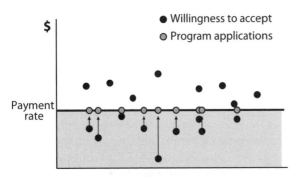

Figure 5. Conservation program applicant pool with practice-based payments

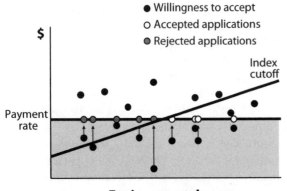

Figure 6. Benefit-cost indices screen applicants

accepted (applications to the right of, or below, the cutoff line in figure 6); others would be rejected (applications to the left of, or above, the cutoff line).

An alternate approach is to specify payment rates based on the environmental score. In figure 7, the cutoff line from figure 6 is used to establish payment rates. I refer to this as a system of "pseudo" performance-based payments because the payment is based on an estimate of performance (provided by the index) rather than actual (monitored) performance. The switch to performance-based payments changes the applicant pool in at least two ways. First, two farms that applied for the uniform practice-based payment no longer apply (see area with horizontal stripes in figure 7). Those farms were rejected for the practice-based payment because they did not offer sufficient environmental gain relative to the cost of the proposed contracts. Their absence from the bid pool reduces administrative burden without reducing the potential environmental cost-effectiveness of the program. Second, two producers on farms who did not apply for the practice-based payment but whose farms offer substantial environmental gain would apply for the performance-based payment (see area with vertical stripes in figure 7). Even though the practice-based payment was insufficient to encourage application, those producers can make a cost-effective contribution to meeting program objectives.

While the switch to performance-based payments has improved the applicant pool, note that there are still a number of farms that could receive payments in excess of their willingness to accept. These surplus payments reduce the funding available to enroll additional producers, reducing the overall cost-effectiveness of the program.

Can bidding stretch program dollars?

Under the right conditions, competitive bidding can broaden the applicant pool and lower payments to individual participants, allowing the enrollment of additional producers within a given budget (Latacz-Lohmann and Van der Hamsvoort, 1997). Suppose the government solicits bids subject to a maximum bid. In figure 8, the bid limit is set uniformly across farms, but

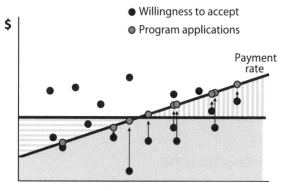

Figure 7. Conservation program applicant pool with (pseudo) performance-based payments

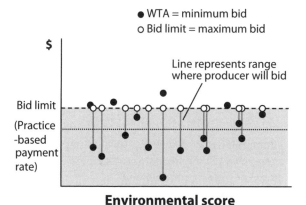

Figure 8. Bid-based payments are between willingness to accept (WTA) and the bid limit

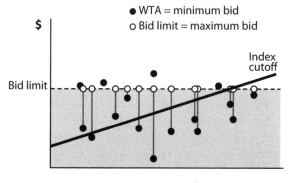

Figure 9. Benefit-cost indices screen applicants with bid-based payments

at a rate higher than the practice-based payment depicted in figures 5 and 6. Producers will bid somewhere between the maximum and their willingness to accept (the farm-specific ranges depicted by vertical lines in figure 8). Which producers are ultimately enrolled depends upon the bids and where the bid cutoff is established (usually the cutoff is established after bids have been received). For the index cutoff that was used in the practice-based payment case, note that the same nine farmers who applied for the performance-based payments could also offer bids at or below the index cutoff (Figure 9). If those producers offer to take payment rates that are below the index cutoff, cost savings could be achieved relative to the performance-based payment. (Also note that the index cutoff would not necessarily be set at this level; the actual level would depend upon the budget and the bids actually received.) If producers do not offer bids at or less than the level of the performance-based payment, however, cost savings would not be achieved.

What conditions encourage producers to reveal their willingness to accept? Competition is critical. If producers believe their offers are unlikely to be rejected, they will be unlikely to offer bids below the maximum rate. Even when competition is substantial, producers may not offer to discount their bids below the maximum (or offer environmentally better but more expensive practices) if they believe their odds of acceptance are good even without offering a discount (or better practices). In particular, if producers feel that they have enough information to assess their chances of enrollment with confidence, they will tailor their bids (in terms of both financial assistance and practices) to achieve an overall environmental benefit-cost score that is just high enough that acceptance is likely.

The government will face a tradeoff in deciding how much information to give producers. On one hand, producers may not recognize how their actions yield environmental gains (or damage), particularly those that occur downstream or downwind (Ribaudo, 2004). If a benefit-cost index is used to rank applications, giving out information on the environmental portion of the index is an effective

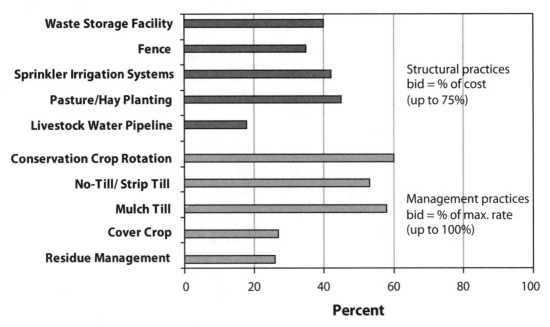

Source: ERS analysis of EQIP contract data.

Figure 10. Average EQIP bids for common practices, 1997-2001

method of conveying information on environmental priorities to producers. On the other hand, producers who recognize that they are offering a field with relatively high environmental potential may be less inclined to offer more effective practices or a lower payment rate if they believe they have a good chance of acceptance without improving their offers. Repeated sign-ups can be problematic because producers may form expectations about acceptable index scores from previous sign-ups—information they can use in calibrating bids to maximize their own return to conservation program participation.

At least two major U.S. conservation programs—CRP and EQIP—have utilized bidding in repeated sign-ups. Between 1997 and 2001, EQIP applicants were asked to bid for financial assistance. During this period, bids were quite low relative to maximum rates (Figure 10). One reason for low bids may have been strong competition. In 1997-1998, only 30 percent of EQIP applications were accepted. Also, many practices commonly funded under EQIP may have produced substantial private benefit. For example, sprinkler irrigation systems that conserve water also reduce

pumping costs; pasture planting can improve grass cover and reduce soil erosion, but will also increase grazing productivity.

In CRP, competitive bidding has been in place since 1991. ("Bidding" in CRP in the late 1980s was not actually competitive bidding because all bids at or below the maximum acceptable rental rate—set for multicounty areas—were accepted.) While CRP applicants are told their EBI environmental scores before finalizing bids, overall EBI scores cannot be determined exactly because the cost-factor weight is set after bids are submitted. Moreover, the EBI cutoff score for a particular sign-up (that determines which contracts will be accepted) is also determined after bids have been received. In practice, however, uncertainty about the cost-factor weight and the EBI cutoff may have been reduced by information from previous sign-ups (see Cason and Gangadharan, 2004, for a discussion of issues related to bidding in sequential sign-ups). In five regular CRP sign-ups held between 1997 and 2003, the EBI changed only slightly: The overall cost weight was 150 points in each of these sign-ups, while the overall EBI cutoff was roughly 200 points in sign-up 15 and hov-

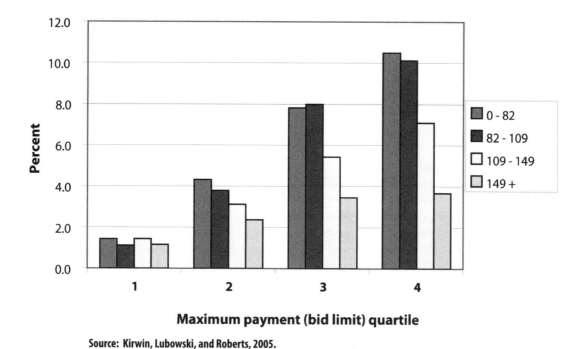

Maximum payment (bid limit) quartile

Source: Kirwin, Lubowski, and Roberts, 2005.

Figure 11. CRP bid discounts as a percent of mean rent offered, general signup 26

ered just under 250 points for sign-ups 16, 18, 20, and 26. (CRP general sign-ups are not numbered sequentially.)

CRP contract offer data, complied by Kirwin et al. (2005), suggests that producers tend to offer smaller discounts when bidding low-cost land (that receives a high-cost score in the EBI) or more expensive land with relatively high environmental potential (which receives a high environmental score). Figure 11 shows bid discounts as a percentage of rents offered for CRP general sign-up 26 in 2003. The lowest value land is in the first quartile by maximum payment (or bid limit), while the highest value land is in the fourth quartile. Bid discounts appear to rise as the value of land rises and EBI cost score declines. The bars represent various levels of exogenous EBI, defined as the EBI score before the producer takes action to improve the score by offering to install better practices or take a lower payment. While the exogenous EBI appears to make little difference to producers offering the lowest value land, producers who offer high-value land with a low exogenous EBI appear to offer larger bid discounts than producers who offer high-value, high-exogenous EBI land.

These results suggest that those who offer land with inherently higher EBI scores also tend to offer less in terms of payment discounts from maximum rates. If so, these producers may be using information gleaned from previous sign-ups to calibrate bids and maximize their own return to CRP participation. While the data shown in Figure 11 are for sign-up 26, the general pattern that emerges from sign-up 26 is consistently observed – to varying degrees – in every CRP general sign-up since 1997 (sign-ups 15 through 26). See Kirwin et al. (2005) for more information.

Summary and conclusions

The level of payment offered to producers for adopting conservation practices is important to the overall cost-effectiveness of conservation programs. On one hand, payments play a key role in determining which producers will apply for participation. Which producers end up in the applicant pool is important to effective targeting through the application of ranking criteria, such as benefit-cost indices. On the other hand, payments that are larger than necessary to secure participation (i.e., larger than a producer's willingness to accept) result in expenditures that could otherwise be devoted to funding additional conservation effort.

Practice-based payments ensure that producers with a low willingness to accept are in the applicant pool, but do not specifically encourage application by producers who can offer greater environmental benefits. Practice-based payments may also exceed the willingness to accept for those who do apply, yielding surplus for individual producers, but limiting the overall number of producers and acres that can be enrolled in the program. Performance-based payments can improve the applicant pool by directing larger incentives to producers who offer greater environmental benefits (and reducing incentives to those who offer little environmental gain). While performance-based payments can help ensure that producers who can offer cost-effective conservation are in the applicant pool, performance payments are also likely to exceed a willingness to accept on some farms, reducing the number of farms and acres that can be enrolled and the overall cost-effectiveness of the program.

A key question is whether competitive bidding can induce producers to reveal their willingness to accept in the context of existing (or similar) conservation programs. What conditions are necessary for producers to reveal their willingness to accept and can those conditions be created in conservation programs? At this time, more research is needed to understand fully problems associated with existing bidding programs and develop alternate bidding (auction) mechanisms that can be effectively applied in conservation programs. Experimental auctions, in which volunteers offer bids under alternate auction mechanisms, have been applied to the problem of conservation program bidding (Cason and Gangadharan, 2004). These methods offer a practical, low-cost way to test auction designs before they are applied in actual programs. Research in this area is ongoing.

Acknowledgements

The author acknowledges the helpful comments of LeRoy Hansen, Mike Roberts, and Marca Weinberg on earlier drafts. Views expressed are those of the author alone and do not necessarily correspond to those of the U.S. Department of Agriculture.

References

Cason, T., and L. Gangadharan. 2004. Auction design for voluntary conservation programs. *American Journal of Agricultural Economics* 86: 1,211-1,217.

Caswell, M., K. Fuglie, C. Ingram, S. Jans, and C. Kascak. 2001. Adoption of agricultural production practices: Lessons learned from the U.S. Department of Agriculture area studies project. Agricultural Economic Report No. 792. U.S. Department of Agriculture, Economic Research Service, Washington, D.C. 110 pp.

Cattaneo, A., R. Claassen, R. Johansson, and M. Weinberg. 2005. Flexible conservation measures on working land: What challenges lie ahead? Economic Research Report No. 5. U.S. Department of Agriculture, Economic Research Service, Washington, D.C. 52 pp.

Claassen, R., A. Cattaneo, and R. Johansson. 2005. Cost-effective design of agri-environmental payment programs: U.S. experience in theory and practice. Paper presented at the Workshop on Payments for Environmental Services in Developed and Developing Countries, Titisee, Germany, June 16-18, 2005.

Cooper, J., and R. Keim. 1996. Incentive payments to encourage farmer adoption of water quality protection practices. *American Journal of Agricultural Economics* 78:54-64.

Khanna, M. 2001. Sequential adoption of site-specific techniques and its implications for nitrogen productivity: A double selectivity model. *American Journal of Agricultural Economics* 83: 35-51.

Kirwin, B., R. Lubowski, and M. Roberts. 2005. How cost-effective are land retirement auctions? Estimating the difference between payments and willingness to accept in the Conservation Reserve Program. *American Journal of Agricultural Economics* 87: 1,239-1,247.

Lambert, D., P. Sullivan, R. Claassen, and L. Foreman. 2006. Conservation-compatible practices and programs: Who participates? Economic Research Report No. 14. U.S. Department of Agriculture, Economic Research Service, Washington, D.C. 43 pp.

Latacz-Lohmann, U., and C. Van der Hamsvoot. 1997. Auctioning conservation contracts: A theoretical analysis and an application. *American Journal of Agricultural Economics* 79: 407-418.

Lichtenberg, E. 2004. Cost-responsiveness of conservation practice adoption: A revealed preference approach. *Journal of Agricultural and Resource Economics* 29: 420-435.

Ribaudo, M.O. 2004. Policy explorations and implications for nonpoint source pollution control: Discussion. *American Journal of Agricultural Economics* 86: 1,220-1,221.

Sandretto, C. 1997. Crop residue management. In Agricultural Resources and Environmental Indicators, 1996-97. Agriculture Handbook No. 705. U.S. Department of Agriculture, Economic Research Service, Washington, D.C. pp. 155-174.

Sheriff, G. 2005. Efficient waste? Why farmers over-apply nutrients and the implications for policy design. *Review of Agricultural Economics* 27: 542-557.

Soule, M.J., A. Tegene, and K.D. Wiebe. 2000. Land tenure and the adoption of conservation practices. *American Journal of Agricultural Economics* 82: 993-1005.

U.S. Department of Agriculture, Farm Service Agency. 2004. Conservation Reserve Program signup 26, Environmental Benefits Index. Washington, D.C.

The disproportionality conundrum

P. Nowak
F.J. Pierce

A conundrum can be a puzzling question posed in the form of a riddle that is answered by a pun, or it can be a difficult problem involving a complex and possibly unknowable solution. Quantifying the environmental benefits of conservation practices on agricultural land at different scales has truly proven to fit this latter definition.

In response to this conundrum, we have introduced a complex array of potential solutions with respect to this issue. None has proved very efficacious, however, in strategically guiding the placement of conservation schemes across the landscape to optimize on both ecological and economic criteria. Our inability to specify the nature, spatial location, and probable impacts of conservation practices on the landscape continues despite spending billions of dollars on what has been called by the chief of the Natural Resources Conservation Service (NRCS) "random acts of conservation."

Driving this large investment has been the creation of what are viewed as innovative programs based on academic theory. Yet the actual impacts of those programs on the environment remain largely unknown. This investment is occurring in an institutional context that bridges the national to the local. Unfortunately, this large cadre of environmental professionals remains focused on implementing programs rather than on solving environmental problems because of our inability to specify where conservation should be placed on the landscape. This is truly a conundrum. Possibly, just possibly, it is time to approach this conundrum from a different perspective.

In developing this alternative perspective, we must begin with fundamentals. Agriculture is essential for human survival, but it is also responsible for natural systems degradation at the global scale (Vitousek et al., 1997; Matson et al., 1997). We know that agricultural landscapes are among the most modified habitats in the world (Robertson, 2000). It is also understood that agriculture plays a major role in changing the processes of biogeochemical systems at varying scales, such as climate (Pielke, 2005; Tilman et al., 2001), species invasions (Blumenthal, 2005), and allocation of scarce water resources within ecosystems (Chakravorty and Zilberman, 2000), among others. Much of this degradation, as noted by Groffman (1997), is due to the fact that agricultural systems are inherently leaky when it comes to impacts on the surrounding ecosystem. Such labels as conventional, sustainable, organic, and holistic are irrelevant when it comes to the basic fact that any agricultural system is inherently leaky. Yet, and this is a critical point, some are more leaky than others. It does not matter if the comparison involves livestock systems, cash-grain systems, horticultural systems, or any of the aforementioned production systems; some agricultural systems are leakier than others. Addressing this simple truism could be a fundamental starting point for determining how to optimize the environmental benefits of conservation practices across agricultural landscapes. An approach to our conundrum may be found in addressing why some agricultural systems are leakier than others.

Any science is concerned with describing, classifying, explaining, predicting, and possibly controlling the variation in the phenomenon studied by that science. Variation is the fundamental subject matter of any science. Without variation there is no need for science. The last paragraph pointed out that there can be as much variation within any classification of agricultural systems as there is between different types of agricultural systems. One would logically expect that conservation programs and the practices they promote would be strongly correlated with the extent of leakage from any population or grouping of agricultural systems. That is, if agricultural systems are inherently leaky, but some are leakier than others, then it is logical to expect conservation to focus on those with the greatest relative degree of leakage.

A major thesis of this chapter, and a possible explanation for the reason that we face the conundrum that we do, is our failure to explore fully this correlation between system leakiness and conservation practice placement. As a result, the description, explanation, and prediction of how and why degradation occurs on agricultural landscapes remains incomplete and, consequently, has a weak relation to placement of conservation practices on the landscape. This missing element in exploring the relation between agricultural degradation and conservation processes relates to the humans who manipulate the biophysical resources associated with any agricultural system (Cabot and Nowak, 2004). Why are some agricultural systems leakier than others? Part of that answer is found in the biophysical setting of that system, and part of it is found in the human dimension that manipulates and manages the system.

The human dimension

We believe that a different perspective can be developed regarding the conundrum of conservation placement in the agricultural landscape by integrating the human dimension with ongoing scientific and policy processes. The difficulty with this objective is that all parties involved in those processes have their own understanding of what constitutes the human dimension. As humans we feel our intuitive or common-sense understanding of humans is all that is necessary and sufficient to guide development and placement of conservation practices. This makes introduction of the human dimension difficult or, as noted by McKelvey and Riddihough:

"In at least one respect, the social, economic, political, and behavioral sciences truly are the "hard" sciences. A problem that is unique to these areas of research is that the subjects of the study (human beings) can read. Because of this, developing a theory to understand and predict an election outcome or a stock market crash is fundamentally more difficult than the problem

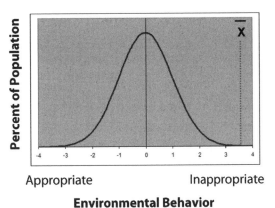

Appropriate   Inappropriate

Environmental Behavior

Figure 1. The Lake Wobegon farmer.

of predicting a chemical reaction or an earthquake" (1999).

Introducing the human dimension is as much a process of debunking existing beliefs and perceptions as it is introducing new insights and perspectives. Humans intuitively understand humans, or least they think they do. For example, a simple question, such as why farmers adopt conservation practices, finds a multitude of answers (Lockertz, 1990). Some of these are disciplinary based (e.g., the economically rational actor), while others are based on anecdotal or informal observations, and still others are based on the logical fallacy of generalizing a few case examples to a larger population. The point is that everyone knows why farmers adopt conservation practices, and any scientific contribution to this query must spend as much time positioning itself in this indigenous knowledge system as it does bringing new scientific findings into the discussion.

What is this prevailing knowledge regarding farmers and conservation? One way to answer this question is through caricature and stereotype. Depending upon who one listens to, there are at least four different types of farmers regarding conservation behavior. The first characterization of farmers relative to conservation behavior might be called "Lake Wobegon" farmers (see Figure 1). All in this group are "above average" when it comes to conservation behavior. While statistically impossible, it is not unusual to hear about farmers collectively characterized as good stewards, the first environmentalists, or members of a group wherein conservation is a way of life.

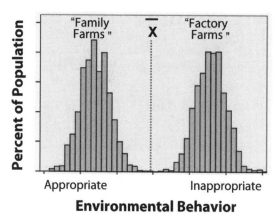

Figure 2. The "us" versus "them" farmer.

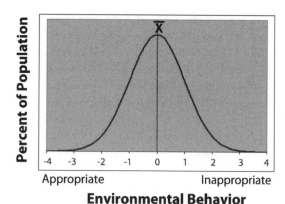

Figure 3. The statistically normal farmer.

From a scientific perspective, this characterization of farmers has little credibility, but it does not stop agricultural organizations or politicians from using it. One has to wonder about the need for targeting of conservation policy with this characterization of farmers.

The second stereotype of farmers is contained in Figure 2, the "us" versus "them" farmers. Again, this unrealistic depiction of our agricultural structure has not prevented its use in polemical arguments. Various types of environmental degradation associated with agricultural production are attributed to "factory farms," despite the lack of comprehensive empirical evidence to support this claim. Family farms, the essence of the agrarian creed in the United States (Dalecki, 1992), are viewed as virtuous, environmentally benign at worst, and threatened by a multinational corporate agriculture. The argument for any conservation policy with this caricature of farmers is based on a regulatory approach focused on "factory" farms, while technical and financial assistance would be focused on "family" farms.

The most common and widely accepted view of farmers is depicted in Figure 3. This is the classic Gaussian distribution where the mean and mode coincide. The expectation associated with this stereotype is that selection of any farmer or farm attribute, especially those associated with conservation behaviors, is normally distributed across the landscape. This depiction of farmers is most commonly found in the various biophysical models used to characterize the environmental impact of agronomic behaviors on the landscape. The various biophysical parameters of these mod-

els are usually measured with rigor and precision, but this stereotype is then used to populate the model with behavioral patterns. For example, soils, slope, hydrology, and other biophysical parameters are measured in association with a specific cropping system. But when there is need to specify the actual behaviors that occur in this well-measured setting, modelers use the recommended or average behaviors associated with nutrient rates, tillage behaviors, or maintenance of conservation practices. In essence, the richness and diversity of farmer behavior across the landscape is reduced to the average, expected, or recommended behavior for a particular setting. Targeting of conservation, consequently, is largely driven by biophysical parameters, with little or no attention to variation in farmer behavior.

The final caricature of farmers is based on a pun associated with George Orwell's classic book *Animal Farm*. In that book one of the lead characters notes that all animals are equal, some are just more equal than others. Thus, the "Orwellian farmer" refers to a situation depicted in Figure 4 and characterized by a log-normal distribution. When examining a population of farmers, therefore, we will find a few saints, a wider array of sinners, and a majority of others falling between these two extremes. Prior research with farmers in Wisconsin, for instance, has found this type of distribution relative to specific, salient resource management behaviors (Nowak et al., 1998; Shepard, 2000). This characterization of farmers is closest to scientific reality, as argued by Limpert (1998) and his colleagues who noted that we live in a "log-normal world." That is, physical, biological,

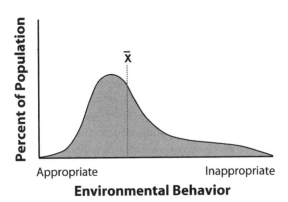

Figure 4. The Orwellian farmer.

and social phenomena are best characterized by a log-normal distribution. The log-normality of our world is an important dimension that has several critical implications for managing agricultural landscapes.

Disproportionality

These critical implications emerge when examining the essence of an agricultural system: The systematic intersection of human behaviors and technologies within biophysical settings to produce food and fiber. As noted earlier, when examining the outcome of these interactions, some are leakier than others. What has been lacking is a more scientifically rigorous depiction of humans in these systems because of the stereotypes and logical fallacies just discussed. Humans, at least from an ecological perspective, are a keystone species. This means that human decisions and resulting behaviors have a major impact on the structure and function of any agricultural system and resulting impact on surrounding ecosystems. Any ecological goods and services, or lack thereof, associated with agricultural systems result from the interaction of humans in a specific biophysical setting. A scientifically rigorous depiction of humans, and not caricature or over-generalization, has to be represented in our analysis of agricultural systems across the landscape.

Why do we still face a conundrum relative to agricultural conservation policy? The answer, we believe, can be found in the truism that some agricultural systems are leakier than others. It follows that conservation policy should be oriented to addressing these leakier systems. The nexus of

the conundrum is associated with our inability to incorporate a scientifically robust human dimension that interacts across diverse biophysical settings in the conservation policy arena. If we accept the lack of validity in the previous caricatures and if we recognize that all farmers do not follow recommended behaviors, then addressing this behavioral variation may be part of the solution to our conundrum.

Because there is variation in both the behavioral and biophysical dimensions of any agricultural system and because science is founded on addressing variation, then the answer to our conundrum will lie in how we characterize this variation. We argue that both the behavioral and the biophysical dimensions of agricultural systems can be characterized by log-normal distributions.

The ability of the biophysical setting to mitigate or minimize the negative impacts of human agronomic behaviors is characterized by a log-normal distribution (Nowak et al., 2006). As noted earlier, studies that have examined populations of farmers relative to those behaviors salient to environmental degradation also found a log-normal distribution. Hence, because any agricultural system results from the interaction of the human and biophysical dimensions, we must examine the resulting interaction of these two log-normal distributions.

The result of this interaction, depicted in Figure 5, presents an answer to why some agricultural systems are leakier than others. Some agricultural systems are leakier than others because of disproportionality. Disproportionality occurs when there is a significant degree of asymmetry between the appropriateness of social behaviors and the buffering capacity, landscape sensitivity, or resiliency of the specific biophysical setting where or when these behaviors take place. Disproportionality has major implications for any targeting scheme. It is a necessary but not sufficient condition to base targeting schemes on biophysical features, such as erodibility of soils, measures of water quality degradation, or habitat suitability, among others. One also needs to understand the nature of the behaviors occurring in those settings. Behaviors appropriate in one setting or time may be inappropriate in others. Note that a value connotation, such as "bad actor," is not used in characterizing this behavior. Instead, the behavior is evaluated

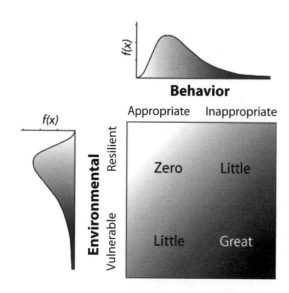

Figure 5. Disproportionality emerges from an interaction of the biophysical and social dimensions of conservation.

on the basis of the best available agricultural and ecological science relative to its appropriateness in a specific setting or time.

This raises a number of challenging scientific questions concerning the scale of targeting. For example, a behavior may be appropriate on a particular portion of a field or during a specific period in a crop cycle, but inappropriate in another portion of that field or during another period in the crop cycle. Other questions relate to how the scale of the technologies employed in agriculture may confound our ability to target remedial actions (Pierce and Nowak, 1999). It is likely that the science behind precision conservation (Berry et al., 2005) may need to find wider applicability in landscapes that are to remain in agriculture. In other settings that are fragile or vulnerable to disturbance, one may need to conclude that there are no appropriate behaviors available, and the land should not be in agricultural production.

The concept of disproportionality is not new to the agricultural sciences that have examined degradation processes. There is ample research that has found common forms of degradation, such as soil erosion (Edwards and Owen, 1991), herbicide losses (Shipitalo and Owen, 2006), and nutrient losses (Sharpley et al., 2001), that occur in disproportionate fashion. That is, an overwhelming

majority of the degradation occurs during a small portion of time or on a small portion of the landscape. The important research that has not been conducted, however, is any measurement of the appropriateness of the behaviors that resulted in these disproportionate contributions.

The prevailing arguments on why we cannot characterize behaviors associated with agricultural systems are associated with costs or confidentiality concerns. But there is a logical sequence of processes associated with using the disproportionality concept in targeting that addresses those concerns. Initially, one needs to characterize the biophysical setting with regard to buffering capacity, landscape sensitivity, or resiliency. At each level of vulnerability, it should be possible to characterize what would constitute appropriate behaviors in those settings. An initial focus on the highly vulnerable settings, however that is measured, should restrict the scope of any targeting initiative to a small proportion of the larger area. For example, nutrient concerns in an agricultural watershed could be restricted to 10 to 15 percent of the spatial area in the watershed by focusing on those areas with the greatest probability of contributing to nutrient runoff. This means that an assessment of the appropriateness of agronomic behaviors only needs to occur with a small proportion of farmers in that watershed. Finally, only a small proportion of those behavioral patterns will be deemed inappropriate and require some form of remedial action. Considering past and current investments in conservation, the relative expense of incorporating the human dimension would represent a very significant gain in efficiency. It is also important to remember that the focus of this effort is to find appropriate behaviors for specific settings or times with a small minority of farmers within the targeted portions of the larger agricultural landscape.

The previous discussion attempted to illustrate that the concept of disproportionality implies that both the biophysical and the social needs to be part of any targeting scheme. It offers a testable hypothesis on why some agricultural systems are leakier than others. While that discussion focused on more applied issues related to designing a targeting scheme, there is also a set of challenging scientific questions associated with the concept of disproportionality. Many of those questions emerge from what may be termed inherent differ-

Table 1. Differences between the biophysical and social dimensions of disproportionality.

Dimension	Biophysical	Social
Space	Spatially heterogeneous	Spatially heterogeneous
Time	Temporally stable	Temporally dynamic
Scale resolution	Fine	Coarse to nonexistent
Policy	Resistant to policy intervention	Conducive to policy interventions
Causality	Lags and feedbacks confound causality	Short-term cause-and-effect outcomes feasible
Surprises	Subject to episodic events and legacy impacts	Subject to episodic events and legacy impacts

ences between the social and biophysical dimensions that would come into consideration when trying to manage agricultural landscapes for environmental quality.

The comparative dimensions of social and biophysical log-normality are summarized in Table 1. Spatial heterogeneity characterizes social and biophysical structures and processes, but the biophysical dimensions are much more stable across time. Resolution of the data, spatial and temporal, available in the biophysical arena, however, is a magnitude of order better than data available in the social area (Goodchild et al., 2000). This leads to the paradox of having more data at a finer resolution on the phenomena that are more difficult to change. That is, regardless of data availability, it is relatively more difficult to change biophysical vulnerabilities on the landscape when compared to changing patterns of human behavior. This results in biophysical features being more resistant to policy interventions, and policy, therefore, must largely focus on social aspects.

A significant amount of research, cited elsewhere in this book, points out how lags in system response and feedback mechanisms create a complex setting for the biogeochemical aspects of ecological systems. While the same complex processes are at work in social systems, again, it is relatively easier to induce cause-and-affect outcomes associated with social processes, at least at finer scales. Finally, both biophysical and social structures and processes are impacted by and often generate episodic events and legacy impacts.

Disproportionality and the future of targeting

We began this discussion by noting that we face a complex and difficult problem for which there are no easy answers. We addressed this conundrum by asking why some agricultural systems are leakier than others. Our answer was based on the concept of disproportionality that emerges from the interaction of the social and biophysical attributes of any agricultural system. Figure 6 represents the targeting strategy that emerges from employing the disproportionality concept to target conservation across agricultural landscapes. By focusing on small portions of the landscape, where inappropriate behaviors are occurring in vulnerable settings, it is possible to derive significant conservation benefits. The magnitude of these benefits declines as one begins to focus on less vulnerable landscapes where more appropriate behaviors are occurring.

Figure 6 also introduces another dimension of the conundrum: Assessing the efficiency and effectiveness of any conservation effort. It is a political reality that any conservation program should be held accountable relative to its efficiency and effectiveness: Efficiency in the sense of the ratio between resources expended versus conservation benefits gained and effectiveness relative to the degree or extent the conservation effort achieves the stated objectives. Any answer to the question of where conservation should be located on the agricultural landscape will be judged by the efficiency and effectiveness of that effort. This is the crux of the Conservation Effects Assessment

Figure 6. A targeting strategy based on disproportionality.

Project, where questions have been raised about whether the efficiency and effectiveness issues should be addressed relative to past and current conservation efforts, or whether it should focus on the future relative to where those efforts should occur (SWCS, 2006).

Beginning to address disproportionality in the agricultural landscape would enhance the efficiency and effectiveness of any conservation effort. It would, however, require a new "mind-set" among the responsible agencies and other partners. Solving problems by working with local farmers to find appropriate behaviors in vulnerable settings is a very different approach than the current method where the focus is on with managing programs where the technical solutions and incentives are determined before behavioral appropriateness is even assessed. We believe that a targeting approach based on the concept of disproportionality is scientifically feasible, but it remains to be seen if it is politically tenable. Not just in terms of our elected decision-makers, but also politically tenable to the administrative and institutional actors who are charged with implementing and monitoring our conservation efforts. Unfortunately, the conundrum remains with us until we are ready to explore the political feasibility of designing a new framework for our conservation programs. Simply generating more science will not address the conundrum. It remains both a scientific and a political challenge.

References

Berry, J., J. Delgado, F. Pierce, and R. Khosla. 2006. Applying spatial analysis for precision conservation across the landscape. *Journal of Soil and Water Conservation* 60(6): 363-370.

Blumenthal, D. 2005. Interrelated causes of plant invasion. *Science* 310: 243-244.

Chakravorty, U., and D. Zilberman. 2000. Management of water resources for agriculture. *Agricultural Economics* 24(1): 3-7.

Dalecki, M.G. 1992. Agrarianism in American society. *Rural Sociology* 57 (1): 48-64.

Edwards, W., and L. Owens. 1991. Large storm effects on total soil erosion. *Journal of Soil and Water Conservation* 46(1): 75-78.

Goodchild M., L. Anselin, R. Applebaum, and B. Harthon. 2000. Towards spatially integrated social science. *International Regional Science Review* 23: 139-159.

Groffman, P. 1997. Ecological constraints on the ability of precision agriculture to improve the environmental performance of agricultural production systems. In J. Lake, G. Bock, and J. Goode, editors, *Precision Agriculture: Spatial and Temporal Variability of Environmental Quality.* John Wiley & Sons, New York, New York. pp. 52-64.

Lockeretz, W. 1990. What have we learned about who conserves soil? *Journal of Soil and Water Consservation* 45, 517-523

Limpert, E., W. Stahel, and M. Abbt. 2001. Log-normal distributions across the sciences: Keys and clues. *BioScience* 51(5): 341–352.

Matson, P., W. Parton, A. Power, and M. Swift. 1997. Agricultural intensification and ecosystem properties. *Science* 277: 504-509.

McKelvey, R., and G. Riddihough. 1999. The hard sciences. *Proceedings of the National Academy of Science* 96(19): 10,549-10,549.

Nowak, P., and P. Cabot. 2004. The human dimensions of resource management. *Journal of Soil and Water Conservation* 59(6): 128-135.

Nowak, P., S. Bowen, and P.E. Cabot. 2006. Disproportionality as a framework for linking social and biophysical systems. *Society and Natural Resources* 19: 153-173.

Pierce, F., and P. Nowak. 1999. Aspects of precision agriculture. *Advances in Agronomy* 67: 1-84.

Pielke, R. 2005. Land use and climate change. *Science* 310: 1,625-1,626.

Robertson, D. 2000. One-dimensional simulation of stratification and dissolved oxygen in McCook reservoir, Illinois. Technical Report 00-4258. U. S. Geological Survey, Reston, Virginia.

Sharpley, A., R. McDowell, J. Weld, and P. Kleinman. 2001. Assessing site vulnerability to phosphorus loss in an agricultural watershed. *Journal of Environmental Quality* 30: 2,026-2,036.

Shipitalo, M., and L. Owen. 2006. Tillage system, application rate, and extreme event effects on herbicide losses in surface runoff. *Journal of Environmental Quality* 35: 2,186-2,194.

Soil and Water Conservation Society. 2006. Final report of the Blue Ribbon Panel conducting an external review of the USDA Conservation Effects Assessment Project. Ankeny, Iowa. http://www.swcs.org/documents/CEAP_Final_Report.pdf

Tilman, D., J. Fargione, B. Wolff, C. D'Antonio, A. Dobson, R. Howarth, D. Schindler, W. Schlesinger, D. Simberloff, and D. Swackhamer. 2001. Forecasting agriculturally driven global environmental change. *Science* 292: 281-284.

Vitousek, P., H. Mooney, J. Lubchenco, and J. Melillo. 1997. Human domination of Earth's ecosystems. *Science* 277: 494-499.

Part 4

Realistic expectations about the timing between conservation implementation and environmental effects

Realistic expectations of timing between conservation and restoration actions and ecological responses

Stan Gregory
Arthur W. Allen
Matthew Baker
Kathryn Boyer
Theo Dillaha
Jane Elliott

Private landowners, citizen groups, local communities, and governmental agencies invest enormous effort, time, land, and money into practices designed to conserve or restore ecosystem functions and structure. A recent survey estimated that river restoration in the United States amounts to more than $1 billion annually (Bernhardt et al., 2005a). In 1995 alone, federal expenditures on watershed-based programs to reduce agricultural pollution were estimated to exceed $500 million (General Accounting Office, 1995). Even though restoration costs are considered high by much of the public and local decision-makers, ecological benefits derived from those efforts are believed to exceed conservation and restoration expenditures (Costanza et al., 1997). For example, a study on a 72-kilometer (45-mile) reach of the Platte River estimated households along the river valued ecosystem services (water quality, soil erosion control, habitat, recreation) delivered at $19 million to $70 million annually, substantially more than the costs of conservation measures undertaken [e.g., water leasing at $1.1 million and Conservation Reserve Program (CRP) contracts of $12.3 million] (Loomis et al., 2000).

The needs, locations, and costs of conservation and restoration are constantly debated—always with passion, sometimes with information. An element frequently missing from these discussions is any realistic estimation of the time required before desired outcomes are attained (Stanford et al., 1996; National Research Council, 2002). While conservation or restoration actions are well-intended, expectations about timing of outcomes and effectiveness of such actions are often unrealistically short. As Wayne Elmore, a rangeland management scientist, noted, "Instant gratification is not fast enough for most Americans." Our objectives here are to identify timeframes over which conservation and restoration outcomes in agriculturally dominated landscapes are likely to be realized; explore landscape, ecological, and social factors affecting the definition of success for these practices; and address how conservation policies can be designed, implemented, and evaluated to yield reasonable measures of the effectiveness of these practices in agricultural ecosystems.

Conservation versus restoration

Conservation and restoration are closely related but distinct processes. Dissimilarity between these concepts has enormous consequences in terms of how success of ecological responses to management actions is defined. Conservation attempts to maintain or protect functional and ecological components of ecosystems to sustain existing resources. In contrast, restoration attempts to repair ecosystem processes and components to restore functions or structure that have been impaired or eliminated. Restoration outcomes range from minor renovation of ecological processes to attempts for complete recovery of ecosystem structure and function, which is rarely attained. Ideally, conservation maintains the performance of the existing system. Depending upon the amount of degradation and degree of recovery possible, restoration may require decades or longer to realize measurable responses. In terms of realistic expectations, one of the most critical distinctions is that conservation attempts to protect existing ecosystem structure and function; desired outcomes can thus be achieved more immediately. A major question that must be addressed is the degree to which responses from these practices can be maintained. In contrast, restoration practices are designed to restore a portion of impaired ecosystem structure and function; thus, desired outcomes may require decades or centuries before restoration goals are realized.

In addition to substantial time lags in ecological responses potentially associated with restoration, the spatial extent and location of restoration may lead to distinct ecosystem responses. As implied by the river continuum concept (Vannote et al., 1980), this is especially true for discharge-dependent characteristics, such as flow regime, water temperature, and water chemistry. The river continuum concept suggests the relative influence of riparian shading and allochthonous inputs should decline as rivers increase in size because (1) channels generally become wider with reduced area of effective shading, (2) the amount of allochthonous riparian carbon is dwarfed by autochthonous in-stream carbon, and (3) increases in the volume of water passing through any particular cross-section require greater inputs of energy or carbon to significantly alter water temperature or allochthonous carbon concentrations. The river continuum concept can be used when scaling expectations of ecosystem response to restoration. For example, measurable impacts of riparian restoration at a given location on water temperature or solute concentrations should only be expected if the restored system shades the channel for a substantial fraction of its sun-exposed length or intercepts a substantial portion of dissolved pollutants. Following this reasoning, the impact of a given restoration effort, as well as the ability to detect effects, depends upon the size of the area (plot, reach, or riparian area) being restored relative to the size of the contributing watershed or stream network. Larger, more spatially complex areas will require greater amounts of restoration effort to achieve similar levels of recovery than can be expected

within smaller areas presenting less physically and ecologically intricate challenges.

Ecological restoration: Successes and failures

Water temperature

Surface water temperature is determined by many variables, but major factors influenced by human activity include water quantity, channel morphology, subsurface exchange, and riparian vegetation (Independent Multidisciplinary Science Team, 2004; Poole and Berman, 2001). Agricultural practices potentially alter all four factors leading to increased rates of thermal alteration (warming and cooling) along stream and river networks. Restoration actions related to water quantity generally focus on reducing withdrawals from surface waters, increasing efficiency of water use, and restoring groundwater sources. Recovery of channel dimensions along streams and rivers in agricultural land commonly requires restoration of riparian plant communities, management of livestock grazing, and reversal of stream channel incision processes. Restoration of subsurface exchange, either hyporheic or groundwater, includes reconnection of hydrologic flow paths (Younus et al., 2000; Ebersole et al., 2003) or restoration of depleted alluvial sediments. Recovery of riparian shade is one of the most common agricultural restoration efforts and includes replanting, natural regeneration, livestock management, and changes in land use (Marsh et al., 2005). The hydrologic, geomorphic, and ecological processes involved in restoration actions require differing amounts of time to achieve their goals. Most require decades at the very least. None can provide immediate recovery of stream temperature and its influence on aquatic ecosystems.

Evidence of temperature response to modification of riparian vegetation in agriculturally dominated basins suggests that removal of riparian vegetation increases stream warming while reestablishment of riparian shade leads to reduced warming (Wehrly et al., 1998; Independent Multidisciplinary Science Team, 2004; Wang et al., 2003). The influence of riparian shade on rates of warming diminishes as streams become wider and discharge increases. A few studies have noted that shade has little or no influence on stream temperature where subsurface inputs are significant (Mosley, 1983), stream water temperature is similar to air temperature (Borman and Larson, 2003), or in large streams where the relative influence of shade on surface water area is minor (Bartholow, 1995). The overwhelming number of studies of wadeable streams, however, concludes that shade influences stream temperature, thus restoration of riparian vegetation may reduce rates of warming and observed stream temperatures (Independent Multidisciplinary Science Team, 2004; Wehrly et al., 2003). Therefore, a portion of stream temperature recovery requires reestablishment of canopy cover over the stream channel. Reestablishment of channel dimensions through riparian recovery may also lead to lower stream temperatures. Reestablishment of effective vegetative canopy cover generally requires 10 to 30 years, depending upon the size of the stream and the type of riparian plant communities restored.

In northern California, late-seral riparian forests maintained summer water temperatures supporting cold-water amphibians and salmonids, while streams in grasslands exhibited higher temperatures (Welsh et al., 2005). Another study in California concluded abundance and distribution of riparian canopy substantially influenced stream temperature in basins up to approximately 75,000 hectares (158,000 acres) (Lewis et al., 2000). Deforestation in Japan resulted in loss of riparian forests and increased maximum temperatures from 22 Celsius degrees (72 Fahrenheit degrees) to 28 Celsius degrees (82 Fahrenheit degrees) during a 50-year period (Nagasaka and Nakamura, 1999). Fish communities in Japan were strongly affected by temperature, with more salmonids in forested reaches than found within grassland reaches (Inouel and Nakano, 2001). Studies in New Zealand observed that removal of riparian vegetation by cattle increased stream temperatures 3.9 Celsius degrees (7.8 Fahrenheit degrees) to 7.8 Celsius degrees (14 Fahrenheit degrees) and altered the macroinvertebrate community structure (Quinn et al., 1992). Investigations of livestock grazing in eastern Oregon found streams with canopy covers greater than 75 percent supported water temperatures meeting thermal requirements for rainbow trout and Chinook salmon. The lowest temperatures were observed in streams without streamside grazing (Maloney et al., 1999). Grass-dominated riparian buffers can provide as much shade as buffers dominated by woody vegetation in small Minnesota streams, but wooded buffers exhibited the lowest maximum stream temperatures (Blann et al., 2002).

Water chemistry

The Chesapeake Bay watershed represents successful coordination among various local, state, and federal agencies, as well as an instructive lesson about expectations from efforts to manage nutrient discharges from urban and agricultural landscapes. Because agriculture is the single greatest source of nutrients in the Chesapeake Bay, significant efforts were directed toward reducing nonpoint-source nutrient inputs into the watershed. Early restoration efforts focused on erosion-based best management practices (BMPs); these were relatively successful at reducing particulate phosphorus losses from agricultural land, but less successful at reducing nitrogen, which is more often transported as dissolved nitrate (Boesch et al., 2001). Most efforts were process-based, however, focusing on landowners developing and implementing nutrient management plans. Reductions in nutrient loads resulting from those plans were typically assumed rather than directly assessed through monitoring of water quality. Although ambitious water quality monitoring programs were able to describe trends at the outlets of major tributaries, it was difficult to discern the causes when restoration activities failed to meet expected objectives. Further analysis suggested increases in annual rainfall during the past decade and time lags associated with dissolved transport in groundwater have occasionally contributed to elevated inputs in surface water, despite improved nonpoint-source nutrient management, further complicating an understanding of restoration efforts (Boesch et al., 2001). Synthesis of results from restoration projects in the Chesapeake Bay watershed (Hasset et al., 2005) suggest that, although the vast majority of restoration has focused on water quality or riparian management, relatively few projects have incorporated follow-up monitoring to assess water quality and ecological benefits. Therefore, it is difficult to evaluate the success of those restorations and their effectiveness in improving water quality in the Chesapeake Bay.

Pollutants and wastes

Characteristics of soils and sediments influence time lags between implementation of management actions and improvement in water quality. If the phosphorus content of soils is high, ceasing the application of manure or fertilizer will eliminate further increases, but many crops must be grown before soil test phosphorus declines to acceptable levels (Read et al., 1973; Halvorson and Black, 1985). As long as phosphorus in soils remains high, the soil will remain a source of particulate and dissolved phosphorous for transport to surface waters. Consequently, the impact of limiting phosphorus applications may not be immediately apparent. Phosphorus can also accumulate in wetland, streambed, and lake sediments. Sediments are a recognized source of phosphorus in the overlying water column and are implicated when the phosphorus content does not decline in proportion to a reduction in inputs (Marsden, 1989). The release of phosphorus from sediments does not occur at a constant rate because of the influence of sediment type, temperature, pH, redox potential, nitrate concentration, and physical disturbance (Holdren and Armstrong, 1980; Jensen and Andersen, 1992). In addition, time lag of ecological response to a conservation measure varies in response to specific environmental conditions. In an example from Washington, Lake Sammamish failed to show an ecological response to a one-third reduction in phosphorus loading for more than 10 years before improving markedly in the subsequent five years (Welsh et al., 1986).

Similarly, soils and sediments can amass pesticides that can contaminate water and impact the ecosystem long after applications have ceased (U.S. Geological Survey, 2006). Continuing detections and impacts of DDT and its metabolites years after discontinuance of use are examples of time lag for response to intervention. Long-term existence of pesticides in stream sediments is greatest for pesticides with little affinity for water (low solubility), but pesticides with relatively high solubility and relatively fast soil degradation rates have also been observed to persist in wetland substrates (Elliott et al., 2001).

Manures applied or excreted on agricultural land contain pathogens that can be transported to surface waters and deposited into sediment (Collins et al., 2005; Muirhead et al., 2006). Although pathogens are not as likely to persist as long as some pesticides, *E. coli* have been observed to survive and even exhibit temporary growth in freshwater sediments in laboratory experiments (LaLiberte and Grimes, 1982). *Escherichia coli* have also been shown to survive for up to six weeks in stream sediments and become resuspended in the water column during storm events (Jamieson et al., 2005).

Veterinary pharmaceuticals are present in manures applied to agricultural land, as witnessed by growing numbers of reports documenting detections of antibiotics and pharmaceuticals in streams and rivers (Koplin et al., 2002; Lindsay et al., 2001) and with clear indications that at least some originate from agricultural operations (Calamari et al., 2003). It is likely that pharmaceuticals will behave similarly to pesticides, possibly remaining is sediments long after their initial introduction to surface water. Diaz-Cruz et al. (2003) reported detections of veterinary drugs in sediments, and Halling-Sorensen et al. (1998) described the presence of persistent antibiotics in sediments of fish-farm sites where antibiotics had been administered.

Given the storage capacity of sediments for nutrients and contaminants, it is unrealistic to expect management alterations that reduce inputs will have an immediate impact on water quality. Even drastic actions, such as the elimination of all pesticide applications, may not reduce concentrations to levels that can be explained by atmospheric transport until the legacy of past pesticide applications remaining in sediments are depleted. Consequently, it is important not to celebrate an apparent success prematurely because pesticides may temporarily disappear from the water column, only to reappear as they are released from sediments (Cessna and Elliott, 2004).

Some management practices have inherent time lags between establishment and their expected environmental response. For example, conservation tillage has been found to reduce soil erosion 70 percent or more in upland areas, but monitoring programs often fail to detect significant reductions in sediment loss at the watershed outlet for a decade or more. This may be the result of a temporary increase in gulley and channel erosion or large quantities of sediment already in storage at the watershed level. Until the channel system reaches a new hydraulic equilibrium with reduced sediment inflows, the sediment that once came from upland areas will be replaced by sediment from channel erosion.

Similarly, establishment of riparian buffers may disturb streambanks and have a temporary negative impact on water quality. Several years may be required for vegetation to become sufficiently established for the buffer to become effective. Conversely, the effectiveness of nutrient removal by an established buffer often declines over time as nutrients accumulate in flow paths (Sheppard et al., 2006).

Climatic effects on hydrology and water quality often have greater effects than could be expected, outweighing environmental responses from a conservation practice (Maulé et al., 2005; Glozier et al., 2006). Simultaneous monitoring of weather and water quality may allow detection of subtle changes due to management that may otherwise be undetectable. Another approach is to examine event hydrographs and only compare pre- and post-management water quality for hydrologically similar events (Glozier et al., 2006). Nonetheless, it should be expected that many years of monitoring data will usually be required to separate conclusively management effects from those affected by climatic variability.

While some ecosystem impairments take decades to recover, other watersheds may respond quickly to conservation measures. Water quality impairments caused by nonconservative contaminants, such as bacteria from human and livestock sources, which die-off or degrade quickly in the environment, have been quickly reduced in some cases. A case in point is the North Fork River total maximum daily loads (TMDL) (U.S. Environmental Protection Agency, 2004). In 1998, a bacteria (fecal coliform) impairment TMDL was developed for the 806-square-kilometer (311-square-mile) watershed in West Virginia. Both point and nonpoint bacteria sources were identified with pastureland, failing septic systems, and direct in-stream deposition via cattle defecation identified as the primary causes of bacterial impairment. The TMDL required a 36 percent load reduction from agriculture and pastureland and no reduction from other sources.

In 1998 the U.S. Department of Agriculture's Natural Resources Conservation Service (NRCS), the Potomac Valley Conservation District, and the

North Fork Watershed Association began work on a management plan to lessen damage from flooding and improve water quality within the watershed. In 2000, the North Fork Watershed Association obtained U.S. Environmental Protection Agency (EPA) 319 funding to implement the management plan. Implemented BMPs included fencing along streambanks, alternative livestock watering facilities, livestock water wells, riparian buffers along streams, nutrient management plans, educational programs, manure and poultry litter composting, and stream restoration. Approximately 85 percent of the farmers in the watershed were actively involved in implementing voluntary, incentive-based BMPs. The North Fork River was delisted for the fecal coliform impairment in 2004, based upon monitoring data collected from 1998-2000 showing that BMPs can effectively address water quality issues.

The effectiveness of riparian buffers, filter strips, and similar practices in reducing pollutant loadings from agricultural land has been heavily researched, but remains poorly understood, with results believed to be extremely site specific. In a review of 72 journal articles, 59 published since 2000, dealing with primary research on the effectiveness of buffers for water quality protection, buffer efficiencies were reported to be relatively high (Table 1).

Unfortunately, reported efficiencies, such as those shown in table 1, may not be representative of real-world buffer efficiencies because most experiments poorly represent field conditions and/or the long-term effectiveness of buffers. Most experimental studies have four serious limitations that constrain effectiveness in representing field conditions:

1. Most buffer research is conducted on small plots constructed so that shallow, uniform flow across the plots is maximized. In the real world, shallow, uniform flow is the exception, and most flow from upland areas crosses buffers as concentrated flow, which greatly reduces buffer effectiveness. Thus, experimental studies that do not consider concentrated flow effects tend to overestimate buffer effectiveness.

2. Most buffer research is conducted on small plots with small source-area-to-buffer-area ratios that are not representative of buffers installed under actual field conditions. For

Table 1. Reported effectiveness of riparian buffers for reducing nonpoint-source pollutants (runoff, sediment, nutrients, and pesticides).

Parameter	Reduction		
	Range (%)	Mean (%)	n
Runoff	21 to 88	51	8
Biological oxygen demand	18	18	1
Ammonium	28 to 87	65	9
Nitrate (runoff)	9 to 99	69	13
Nitrate (subsurface)	49 to 91	72	6
Phosphate	36 to 98	73	8
Total Kjeldahl nitrogen	11 to 79	48	5
Total nitrogen	37 to 94	64	11
Total phosphorus	5 to 91	61	18
Sediment	17 to 100	84	69
Atrazine	22 to 70	51	6
Metolachlor	51 to 66	56	3
Fecal coliform	28 to 90	67	4

example, across the 69 studies analyzed in table 1, the source-to-buffer area ratio ranged from 0.4:1 to 55:1, with a median of 5.5:1. This median value would require the conversion of 18 percent of agricultural land to buffers and is considerably higher (two to three times) than the recommended, or allowed, ratio in most buffer programs.

3. Most experimental buffer studies are conducted on newly established buffers (typically less than a year since establishment), with most monitoring lasting for less than a month. Thus, most experimental buffer study results represent effectiveness only during establishment, failing to furnish estimates of long-term effectiveness.

4. Most experimental buffer studies on cropland either use or simulate conventional tillage and agrochemical applications in the experimental area. This is unrealistic. Buffers should only be used in concert with in-field systems of conservation practices designed to keep sediment and agricultural chemicals in the field where they are valuable resources rather than pollutants that need to be trapped

by buffers (Dillaha et al., 1989). The few buffer experiments simulating high sediment and nutrient loadings over longer time periods suggest the effectiveness of overloaded buffers will decline dramatically over time.

Only one of the studies summarized in table 1 (Udawatta et al., 2002) simulated real world conditions in terms of concentrated flow patterns, reasonable source-to-buffer-area ratio, and use of infield conservation practices (no-till) in addition to buffers. This three-year study used a paired approach, with a control watershed in row crops [1.6 hectares (4 acres)] and two treatment watersheds: One with grass buffer strips [3.2 hectares (7.9 acres)] and the other with trees in grass buffer strips [4.5 hectares (11 acres)]. No-till was used on cropland in all three watersheds. Grass buffers and trees in the agroforestry treatment were established in 1997, with monitoring initiated at the same time. The buffers consisted of a system of in-field contour buffers and grass waterways along major in-field drainageways. The cropland-to-buffer-area ratio was approximately 8:1, with about 13 percent of the treatment watershed area devoted to buffers. Runoff, sediment, and nutrient losses were monitored at watershed outlets. The control watershed had total phosphorus, total nitrogen, and nitrate losses of 0.42, 1.52, and 0.28 kilograms per hectare per year (0.92, 3.36, and 0.63 pounds per acre per year), respectively, indicating no-till was effective in minimizing nutrient losses without buffers. The grass buffer and agroforestry treatments reduced surface runoff 10 percent and 1 percent, respectively; sediment losses increased 35 percent and 17 percent, respectively; total phosphorus losses declined 8 percent and 17 percent, respectively; total nitrogen losses declined 14 percent and 11 percent, respectively; and nitrate losses declined 21 percent and 5 percent, respectively. The reported increases in sediment losses with the buffers were unexpected, but the losses with the source areas in no-till were so low that the increase was negligible. Sediment losses from the control, grass buffer, and agro-forestry treatments were 27, 33, and 36 kilograms per hectare per year (60, 72, and 79 pounds per acre per year) over the three-year study, which is extremely low and indicative of excellent no-till production.

Aquatic communities

The extent of actions intended to improve aquatic habitats in agricultural landscapes varies across the United States and Canada because agricultural land is generally privately owned and management objectives may not include concern for fish and wildlife habitats. While U.S. Department of Agriculture (USDA) farm bill programs offer increasingly attractive financial incentives for conservation of aquatic resources, the degree to which restorative actions are implemented and monitored for effectiveness is challenging to evaluate and report. This is apparent by the poor rate at which restoration projects have been evaluated (Bernhardt et al., 2005a). This lack of evaluation is a consequence of limited dollars allocated for monitoring and a failure by those who formulate conservation policies to recognize the importance of long-term monitoring to refine performance of conservation programs and practices. Monitoring designs are necessarily intricate and expensive to implement because of the ecologically complex nature of stream, river, floodplain, and upland processes. Nevertheless, monitoring is essential to determine what works and does not work under different circumstances and to gain knowledge

on how long it takes for conservation practices to become effective. In a blue ribbon panel's review of USDA's Conservation Effects Assessment Project (CEAP) (Soil and Water Conservation Society, 2006), panel members concluded that lack of resources for monitoring was a significant limitation of CEAP and other conservation programs. They concluded: "The most important and troubling missing piece is the absence of plans for on-the-ground monitoring of change in the environmental indicators and outcomes conservation programs and activities are intended to improve." The panel recommended that Congress mandate that at least one percent of the funding for each authorized program—about $40 million of the $4 billion U.S. taxpayers are investing in conservation— be set aside to support monitoring and evaluation of those programs.

Restoration actions targeted to improve habitats for aquatic species are difficult to evaluate because effects can be influenced by physical, biological, and chemical responses at multiple spatial and temporal scales having variable affects on biological communities (Minns et al., 1996; Lammert and Allan, 1999; Fitzpatrick et al., 2001; Vondracek et al., 2005). Moreover, suites of practices installed either sporadically or strategically in a catchment will differentially influence the breadth and timing of response of stream or wetland species and their physical habitats. Thus, correlations between a specific practice and the ecological response of an organism or its habitat are not easily discerned. These limitations aside, recent studies focusing on effects of agricultural practices on conservation of aquatic species and their habitats are beginning to offer insights into which practices may be effective at arresting declines in North American aquatic species. In most cases, management practices that retain or improve connections among ecological processes and/or different aquatic habitats contribute to the quality of those habitats and the well-being of the aquatic species that inhabit them.

Along stream and river corridors, fish, amphibians, and aquatic insects move among different habitat types, including pools, riffles, backwaters, wetlands, sloughs, alcoves, hyporheic zones, and riparian zones during their life cycles. Agricultural practices can be modified to maintain connections between essential components of habitat across space and time. In 20 streams in agricultural land within the Minnesota River Basin, wooded riparian areas supported higher fish richness, diversity, indices of biotic integrity, and macroinvertebrate communities than recorded within nonwooded, open reaches (Stauffer et al., 2000). Restoration practices that effectively reconnect upstream and downstream aquatic habitats include providing fish passage around or through barriers, such as dams or poorly constructed culverts (Pess et al., 1998; Hart et al., 2002; Johnson, 2002). Breaching dikes potentially reconnects riverine migration routes with estuarine rearing and holding habitats (Frenkel and Morlan, 1991). Installation and active management of water control structures in constructed or restored wetlands have been effective in preventing entrapment, allowing fish to emigrate out of floodplain wetlands entered during seasonal high flows (Swales and Levings, 1989, Thomson et al., 2005; Henning, 2005).

Keeping fish and water in streams and out of irrigation ditches increasingly is an objective of ranchers and farmers in the arid west, triggering installation of sophisticated fish screens for irrigation diversions (McMichael et al., 2004) and effective irrigation conservation management techniques through candidate conservation agreements [David Smith, USDA-NRCS, personal communication: (http://www.mt.nrcs.usda.gov/about/mtstcm/feb05/grayling.html).]

Simply maintaining physical connectivity between intermittent stream channels used as drainage ditches and mainstem rivers has been shown to influence the amount of winter habitat for native fish, benthic invertebrates, and amphibian species in the grass seed farms of the Willamette Valley of Oregon (Colvin, 2005). Similarly, maintaining open drains on agricultural land in Ontario provides habitat for fish assemblages identical to those inhabiting nearby streams (Stammler, 2005).

Connecting habitats includes maintaining ecological linkages between riparian zones and streams. For example, riparian vegetation structure influences the composition and abundance of terrestrial insect communities. By altering grazing management regimes to favor persistence of riparian vegetation where terrestrial insects thrive, fish benefit from seasonally important food sources. Grazing systems that allow cattle to graze for short durations increase terrestrial insect production, which has been shown to correlate

strongly with fish condition and survival on Wyoming ranchland (Saunders, 2006; Saunders and Fausch, 2006).

Loss of cropland due to streambank erosion has elevated interest in riparian management that includes replanting of herbaceous and woody riparian buffers, often coupled with instream rock or wood to deflect the flow away from unprotected banks. Preliminary investigations in western Oregon indicate such streambank stabilization practices, if designed correctly, encourage instream processes important to aquatic species, including retention of detritus and large wood for fish cover and macroinvertebrate food sources (Stan Gregory, unpublished data). Studies in Minnesota further support the importance of riparian corridor conservation and restoration to aquatic species because it contributes to instream habitat and geomorphic features at multiple scales (Stauffer et al., 2000; Blann et al., 2002; Talmage et al., 2002).

Instream structural improvements have improved fish habitats at some sites. Assessment of the effectiveness of instream structures placed in western Washington and Oregon streams over the last three decades revealed that the majority of sites exhibited significantly higher densities of juvenile coho salmon, steelhead, and cutthroat trout after restoration (Roni and Quinn, 2001). While placement of instream log structures has proven valuable in the Northwest, failures in the effectiveness of this practice in the southeastern United States indicate re-introduction of large wood to drastically altered stream systems is often unsuccessful when placed in stream reaches physically unable to retain them (Shields et al., 2006).

Terrestrial wildlife

The purpose of USDA conservation programs is not to restore native ecosystems but to lessen undesirable environmental effects of agricultural production. Yet these policies, at times imperfect, have brought about significant improvement in the quality and distribution of wildlife habitats associated with agricultural land across much of the American landscape. Fundamental to the design of successful conservation and agricultural policies is recognition that farming and environmental quality improvements are not mutually exclusive goals, nor are environmental solutions associated with soil erosion, water quality, and wildlife habitats independent issues.

Established in 1986, embedded within all 50 states, and composed of an eclectic mix of conservation practices, the 14.6-million-hectare (36-million-acre-plus) CRP represents a cornerstone of USDA conservation policy. Investigations describing the environmental, social, and economic effects of CRP offer insight on at least some effects of conservation policies on wildlife and their habitats (Allen and Vandever, 2005; Haufler, 2005). Some benefits have been profound, such as 25 million ducks produced in the Prairie Pothole region due to the nesting cover provided by CRP grassland. Other benefits are more understated—doubling of the range of mule deer across the Texas Panhandle, for example, or the reversal in population declines of various songbird species in response to CRP grassland replacing crops on highly erodible land. Many CRP conservation practices (e.g., planting of native and introduced grasses, field borders, riparian buffers) are implemented in other federal and state conservation programs. It seems reasonable to assume wildlife-related effects described for individual CRP conservation practices have similar benefits and consequences when applied as part of these other programs as well.

Economic and social support for rural communities, aesthetically pleasing landscapes, recreational opportunities, and sustainable populations of wildlife represent ecosystem services delivered from agricultural land use whose importance is not often adequately captured in assessments (Feather et al., 1999; Costanza et al., 2000). Although wide-ranging personal and social effects of the CRP remain impractical to measure, these nonquantifiable benefits are valued particularly by those most directly affected. CRP participants attribute improving future productivity of land, retention of water from rain and snow, reappearance of springs, improved quality of well water, prevention of unwanted urban expansion, stability in income, lower operational costs, and control of drifting snow as program benefits (Johnson and Maxwell, 2001; Bangsund et al., 2002; Allen and Vandever, 2003). For many, the CRP has enhanced aesthetic qualities of their farmland, brought greater numbers of wildlife, and increased opportunities for recreational and social use of their land. Many of these benefits were delivered soon after establishment of con-servation practices, but an accurate assessment of their economic and social significance remains elusive.

For the sake of simplicity, visualize most wildlife inhabiting agriculturally dominated regions as belonging in one of two groups. Farmland wildlife (e.g., ring-necked pheasant, bobwhite quail, white-tailed deer) generally prosper where a relatively small proportion (e.g., less than 10 percent) of the landscape is dedicated to nonfarmed vegetation, with crop production remaining the prevailing land use. The other category can be characterized as wildlife endemic to grassland (e.g., upland nesting waterfowl, prairie chickens, and pronghorn antelope). These species are generally dependent upon relatively large, contiguous blocks of grassland cover. Farmland species benefit from high levels of interspersion between farmed and nonfarmed land uses; most wildlife species endemic to grassland ecosystems do not.

Conservation programs administered by USDA have benefited species whose elemental habitat requirements are met by conservation practices designed to most appropriately address regionally prevalent forces of soil erosion. In drier, western regions, whole fields planted to grasses offer the greatest opportunities to address wind erosion and the needs of grassland wildlife. In wetter climates, where soil erosion by water is an issue of greater concern, grass filter strips, riparian buffers, field borders, and removal of smaller tracts of erodible land from cultivation typically enhance habitat quality for farmland wildlife adapted to higher levels of interspersion between land uses.

Time lags in ecological responses

Restoration practices inherently require variable periods of time for ecological processes to deliver desired outcomes, for systems to adjust to restoration measures, for invasive species to be reduced, for desirable endemic species to increase, for toxicants and other forms of degradation to be eliminated or isolated in long-term storage, and for connections between habitats, communities, and ecosystems to be restored (Harding et al., 1998; Sarr, 2002; Bond and Lake, 2003). Wetland restoration studies in the southeastern United States found recovery of amphibian communities was affected by drought and disease after seven breeding seasons, leading to the conclusion that long timeframes are necessary for monitoring programs to assess accurately the outcomes of restoration practices (Petranka et al., 2003). Time lags in replanted vegetation reaching maturity were identified as one of the most serious limitations of restoration for birds and arboreal mammals in Australian agricultural landscapes (Vesk and MacNally, 2006). Monitoring vegetative characteristics of CRP grassland in the Great Plains over a 12-year period, Cade et al. (2004) found that vegetative variables affecting the quality of wildlife habitats varied not only by grass species planted, but also through time and in response to natural or human-induced disturbance. Harding et al. (1998) concluded that the best predictors of present macroinvertebrate communities in streams of the southeastern United States were land use and land cover conditions in the 1950s. The influences of past agricultural land uses on invertebrate communities were still evident after more than 45 years, even though the local riparian areas had become reforested. As Bond and Lake (2003) noted, "…legacies of past disturbances and the impacts of on-going disturbances operating at larger (possibly catchment-wide) scales can compromise works done at individual sites or reaches."

Temperature

Restoration of thermal regimes in stream networks is dependent upon processes that influence shade, discharge, channel dimension, and hyporheic exchange. Restoration of riparian shade clearly requires many years for an adequate, contiguous vegetative canopy to develop along a reach. Geomorphic processes may require decades to adjust channel dimensions, and reconnection of hydrologic flow paths for subsurface exchange are functions of channel structure and hydrologic regimes. A New Zealand study compared physical and biological characteristics of nine riparian buffers, replanted and fenced between 2 and 24 years, with conditions found within control reaches (Parkyn et al., 2003). Some stream properties, such as water clarity and channel stability within treated reaches, responded rapidly. Other characteristics, such as nutrient concentrations and presence of fecal coliform bacteria, were highly variable. Macroinvertebrate community composition did not respond within the time period investigated, which was attrib-

uted to the lack of response in stream temperature. Stream temperature could not be expected to adjust until canopy cover by riparian vegetation had recovered (Quinn et al., 1992).

Past or future changes in hydrologic connections can affect the location and timing of thermal responses to restoration. Roads, ditches, and diversions can also influence stream temperatures by changing the routing of surface and subsurface flows, which may be warmer or cooler than the stream temperature (Story et al., 2003). Consequently, stream temperatures may not respond to recovery of riparian vegetation if the routing of water from ditches or drains significantly alters stream temperatures. Also, restoration of hydrologic connectivity and detention through recovery of hyporheic zones through channel aggradation or reestablishment of wetlands may require several years or decades for hydrologic paths to become reestablished and well integrated into the flow network.

Nutrients and contaminants

Groundwater nitrate, leached from surface soils via subsurface flow to near-stream zones, may be an important source of nitrogen to surface waters (Cirmo and McDonnell, 1997). In some river systems, groundwater can make up as much as 50 percent of river flow, and groundwater may be decades to centuries older than surface water (Michel, 1992). In such systems the potential for time lags in water delivery can have a profound impact on the ability to detect degraded systems and quantitatively describe responses to restoration. For example, if recent land use practices lead to eutrophication of surface waters, it is possible for dilution by older and deeper flow systems with higher quality water to mitigate observed water quality degradation, particularly during baseflow conditions. On the other hand, shallow groundwater can retain nitrate concentrations for 40 years or more in the absence of reducing sediments (Bohlke and Denver, 1995). In these systems, detecting positive effects in post-restoration monitoring can be hampered by delivery of enriched, pre-restoration water to the stream. In such flow systems, however, long time lags reflect slow rates of delivery; hence, the ability of deeper flow systems to influence instantaneous stream concentrations would require substantial groundwater sourcing. Thus, although the full benefits of restoration practices may be masked in some systems by lags imposed on nitrogen-enriched groundwater, significant masking after a decade should be unusual.

In Mid-Atlantic states, nitrogen leaching from tributary watersheds of the Chesapeake Bay has increased since 1985 despite widespread restoration activity (Lindsey et al., 2003). Although patterns of individual watershed discharges vary, there is no clear trend across the basin (Alexander and Smith, 2006), leading to concerns about the effectiveness of nearly 20 years of restoration efforts under Chesapeake Bay agreements (Boesch et al., 2001). One recent study showed that although base flow is made up of water between 1 and 50 years old most water in the Chesapeake Bay watershed enters streams within a decade (Lindsey et al., 2003). Although the proportion of baseflow in streams can be influenced by the quantity of annual precipitation, average residence times for groundwater range from 10 to 20 years (Michel, 1992; Focazio et al., 1997). A comparable range of 2 to 9 years has been observed for nitrogen concentrations in waters at the Mississippi River outlet (McIssac et al., 2001), as well as 5 to 10 years for large rivers in Latvia (Stalnacke et al., 2003).

The time lag in nitrogen recovery introduced through soil percolation and groundwater contribution to surface water is relatively short compared to the lag expected in phosphorus recovery due to percolation pathways (Oenema and Roest, 1998). In soils with low phosphorus sorption capacities, unsustainable additions lead to soil saturation, and thereafter phosphorus concentrations in groundwater will increase with the degree of phosphorus saturation. Under these conditions, conservation actions that act to reduce or eliminate phosphorus application surpluses will have no immediate impact on phosphorus reaching surface waters by the percolation pathway. Model estimates suggest phosphorus transport through surface pathways may respond within 5 to 50 years, but phosphorus moving by the percolation pathway may take centuries to respond to management changes (Schippers et al., 2006).

Besides limiting observed benefits of restoration, knowledge of subsurface flow pathways can increase understanding about effectiveness of restoration activities. Molenat and Gascuel-Odoux (2002) showed that reduced nitrogen leaching

along a 500-meter (547-yard) field-to-stream transect with three distinct flow pathways lowered recharge nitrogen concentrations from 100 to 80 milligrams per liter (100 to 80 parts per million) while simulated stream concentrations declined from 57.4 to 45.9 milligrams per liter. Water lag times in this study ranged from less than one year to three years. By redistributing patterns of nitrogen leaching to take advantage of longer travel times and denitrification from pyrite-rich subsurface sediment layers, the authors achieved similar reductions in simulated stream concentrations without changing average groundwater loadings. Thus, in addition to clarifying understanding about the timing of restoration effects, knowledge of groundwater flow pathways can be used as a mitigation or restoration tool to help reduce stream nutrient concentrations (Lindsey et al., 2003).

The Walnut Creek monitoring project in central Iowa investigated response of stream nitrate concentrations to changing land use patterns in a 5,218-hectare (12,894-acre) agricultural watershed over 10 years (Schilling and Spooner, 2006). In 1990, soybeans and corn constituted 69 percent of land use in the Walnut Creek watershed. Between 1990 and 2005, land devoted to row crops declined from 69 percent to 54 percent of the watershed area as a consequence of a U.S. Fish and Wildlife Service prairie restoration project. As a result of the land use changes and implementation of nutrient management programs between 1995 and 2005, nitrogen applications in the watershed declined 21 percent. Nitrate concentrations, however, still exceeded the standard of 10 milligrams of nitrate-nitrogen per liter for drinking water, with concentrations highest in the spring and early summer. Over the 10-year monitoring period, trend analysis indicated nitrate concentrations declined by about 0.12 milligrams per liter per year, or a total of 1.2 milligrams per liter for the whole basin, and by 8 to 12 milligrams per liter in smaller subbasins if a control watershed was used as a covariate. Without adjusting for the control, the reduction was 0.07 milligram per liter per year for the overall basin. Schilling and Spooner (2006) had estimated that a 10 percent change in row-crop area was required for a 1.95-milligrams-per-liter change in nitrate levels over a 10-year period. The lag time between reduced applications of nitrogen fertilizer and nitrate levels in Walnut Creek was influenced by the mean residence time for groundwater, which was estimated to be 14 years. Consequently, Schilling and Spooner (2006) concluded that it was impractical to detect changes in nitrate water quality in larger watersheds in less than several decades, and documentation of improvements in water quality due to conservation practices should focus on small subbasins where changes can be detected in shorter time frames.

Another mechanism influencing efficiency of denitrification in riparian areas is hydrologic connection between enriched groundwater and biogeochemically active sediments (Hill, 1996). Results from investigations in a series of European riparian areas suggested differences as small as 20 to 30 centimeters (8 to 12 inches) in water table depth had a significant effect on denitrification rates (Hefting et al., 2004). Channel incision and/or ditching to improve field drainage are common in agricultural land use, though sometimes incision is an unintended consequence of increasing channel flows. Such hydrologic modification can result in disconnection between enriched groundwater and denitrifying soil layers. Thus, restoration success can be hampered both by changes in hydrologic routing that reduce exposure of nitrogen-enriched waters to denitrifying sediments and alteration of the redox conditions required for denitrification (Pinay et al., 2002). Across whole watersheds, lack of hydrologic connection between nutrient sources and streams can lead to poorly buffered systems, even when a substantial portion of near-stream zones are forested (Weller et al., 1998; Baker et al., 2006).

Aquatic communities

The challenges of detecting and describing ecological successes or failures in improving conditions for aquatic species are due to multiple factors, not the least of which is the inherent variability in life-history patterns of aquatic species. Because fish assemblages are variable from day to day, month to month, year to year, and longer periods, data collected at randomly selected sites to determine if fish are responding to habitat improvements are difficult to interpret (Adams et al., 2004). This challenge may, however, be less daunting than the conflict between time lags in responses of species, habitats, and landscapes and the essentially nonecological timeframes of

human systems. Farm policy, political administrations, landowner dynamics, and agency personnel change many times before watersheds can demonstrate recovery. Legislators want proof that restoration actions are worth the money invested, yet scientists provide only scant amounts of data that often cannot unequivocally prove success in the timeframe demanded by those who formulate or fund legislation affecting conservation policies. Failure to recognize complexities of natural and managed systems, recognition of time lags after implementation of conservation practices, and the historical lack of funding in support of long-term monitoring programs are underlying causes limiting the ability of science to answer fundamental questions about effectiveness of conservation practices and policies on aquatic species. Dynamic systems, such as rivers and streams, change constantly in response to natural disturbances and human perturbations. Conservation policymakers need to recognize change is not only normal in ecological systems, but confounding. Existing environmental issues and unanticipated effects of land use have occurred over decades and centuries. In most cases, it is unreasonable to expect that conservation or restoration will have immediate and permanent benefits to aquatic species and their habitats.

On the other hand, it is quite reasonable to assume that changes in land use practices in uplands will influence the habitats of aquatic species because aquatic systems are a reflection of environmental conditions in a watershed. Conservation tillage, residue management, and conservation buffers that improve overall surface water quality will, over time, benefit the species that use surface waters as habitats. Similarly, where clear, cold water exists, coldwater species can likely exist. Thus, conditions that influence temporal changes in stream temperature (as described previously) also influence temporal species responses. Conservation practices, such as riparian buffers designed to provide shade and channel features that maintain coolwater refuges, will over time provide habitat for species in search of such habitats, assuming a population source exists and barriers do not restrain immigration to those habitats. Some restoration measures do result in an immediate response by fish. Studies in the Pacific Northwest demonstrate success in reconnecting migratory routes and their habitats for anadromous salmonids (Beamer et al. 1998) and providing cover (Roni and Quinn, 2001). Kanehl et al. (1997) evaluated removal of a low-head dam and determined that both stream habitat and desired fish assemblage improved within five years.

Terrestrial wildlife

Effects of conservation polices on wildlife may be seen in a relatively short period of time or may take years to yield observable results. Removal of environmentally sensitive land from crop production has brought observable and immediate benefits to some species, but effects of alternative production and conservation practices, such as minimum tillage and terraces, are not always obvious or quantifiable. The cumulative effects of these practices, however, contribute to improvements in the quality of aquatic habitats downstream from the fields where the practices are applied.

A majority of investigations describing CRP effects on wildlife and their habitats have been completed on the scale of individual fields or by conservation practice (e.g., riparian buffers). The presence of conservation features in isolation, however, rarely has a definitive influence on abundance and distribution of many wildlife species. Rather, overall land use, cropping practices, and the spatial configuration of conservation practices with land remaining in production define long-term capabilities of agriculturally dominated landscapes to support viable populations of wildlife (Rodgers, 1999; Krapu et al., 2004; Taylor et al., 2006). Specifically linking quantitative responses of wildlife with conservation practices depends upon the species in question and becomes complex because wildlife species respond differently as vegetative characteristics change through time and in response to application, or absence, of disturbance brought on by tillage, fire, grazing, or other management practices (McCoy et al., 2001; Fritcher et al., 2004; Cade et al., 2005). Individual conservation practices may be beneficial for one species, but have negative effects on the suitability of habitat for others. For example, in the Texas panhandle, mule deer have expanded their range into heavily farmed landscapes as a consequence of the cover provided by introduced species of grass under the CRP. The same conservation practice, however, has con-

currently diminished availability of habitat for swift fox because the vegetation becomes too tall and unsuitable for the animal's use (Kamler et al., 2001; Kamler et al., 2003).

During the past two decades, there have been many outstanding studies on how wildlife responds to the inclusion of conservation practices in intensively farmed landscapes. These investigations have been, and continue to be, used to refine USDA conservation policies and management guidelines. Hard numbers or measures are needed through which progress toward specific goals can be measured. Wildlife management in agricultural landscapes is well described; however, it is difficult to predict how numbers or distributions of wildlife will change in response to conservation practices. The one overarching criticism that might be directed toward research into wildlife response to conservation policies within agricultural ecosystems is a lack of focus on specific species, making identification of precise, quantifiable goals difficult. If specific goals cannot be identified for unique areas (e.g., farm, watershed, region) it is impossible to furnish measures that accurately describe progress toward reaching those goals.

Wildlife response to contemporary conservation policies in agricultural landscapes is potentially diverse, but it is not possible to optimize management for all species. There are wildlife species whose abundance and distribution reflect a practical balance between conservation and economically viable agriculture. Across much of the Great Plains and Corn Belt, for example, the ring-necked pheasant is perceived as a symbol of balance between agricultural production, conservation, and social value. The same circumstance is represented by upland-nesting waterfowl in the northern Great Plains, across the Southeast by the bobwhite quail, and by anadromous fisheries and sage grouse in the Northwest. Grassland birds in the Northeast are also species that can stand as emblems of balance between agriculture and conservation. These are generally the species about whose habitat needs the most is known. If habitat for these species is furnished, the needs for many, not all, other wildlife species inhabiting agriculturally dominated landscapes will be provided. It is the known habitat needs of these species defined at the field, farm, and watershed levels that offer greatest potential to define beneficial management practices and measurable goals through which the effectiveness of conservation can be more precisely described.

Acceptance of conservation goals affecting wildlife habitat and environmental quality in agricultural landscapes presents social as well as scientific challenges. Conservation programs have been an important source of income for small, intermediate, and rural-residence landowners who are less likely to adopt practices requiring substantial economic investment, technical skills, or management-intensive alternatives (Lambert et al., 2006). Larger operators, whose primary occupation is farming, are more likely to dedicate a smaller percentage of their land to conservation, but they are more likely to install practices generally requiring higher costs and compatibility with sustainable production of crops. The desires and limitations of landowners with differing personal and economic goals must be a part of any successful effort to enhance wildlife habitats associated with agricultural land use over the long-term.

Measuring cumulative effects

In many ways, "cumulative effects" is a vague concept applied to complex interactions. Rigorous scientific assessment of cumulative effects most commonly addresses coupled processes that lead to complex outcomes, but often does not fully address the full range of collective effects. In many ways, the spatial, temporal, and social complexity of landscape-level cumulative effects far exceeds the capacity of most environmental measurement and analysis systems. Yet management of simple sets of processes or small numbers of target species often leads to overly simplistic conclusions and adoption of practices that may degrade other resources. Analyses of multiple factors and processes along river networks has provided important frameworks for restoration of stream ecosystems and associated riparian areas (Li et al., 1994; Gore and Shields 1995) that may be applicable for evaluation of other conservation and restoration practices within agriculturally dominated landscapes.

Cumulative effects of riparian buffers and nutrient responses

Although much effort has been focused on the benefit of riparian buffers and restoration at local sites, comparatively little work has addressed cumulative downstream impacts on water quality (Dosskey, 2001). Recent advances in use of stable isotopes seem promising (e.g., Bohlke et al., 2004), but few tools exist to distinguish permanent from temporary nitrogen sinks across whole watersheds and signal a definitive response to restoration. Because most agricultural land use patterns reflect aggregate land use decisions by individual landowners and most watercourses within watersheds are not well-buffered, it is difficult to detect and measure effects of restoration activities. Baker et al. (in press) recently studied land-cover patterns in more than 500 watersheds from four physiographic provinces within the Chesapeake Bay watershed. The authors compared watershed cropland proportions with proportions adjusted downward to represent presumed effects of existing riparian forests and wetlands. In this manner, they sought to examine whether extant patterns of riparian buffers were likely to result in reduced nutrient discharges compared to those expected from unbuffered areas. Results of the investigation led the authors to conclude that even when riparian buffers were assumed to reduce nutrient concentrations as effectively as in published studies (e.g., Lowrance et al., 1997; also see summary of studies in Table 1) most watersheds showed buffer patterns that would not lead to a substantial reduction in nutrient discharges. This finding underscores the need to design riparian buffer restoration to intercept and process multiple upslope sources and coordinate with other watershed strategies to achieve intended outcomes.

Most studies of riparian buffers demonstrate water quality benefits measured along field-to-stream transects (e.g., Peterjohn and Correll 1984; Lowrance et al., 1997) or describe substantial denitrification potential (e.g., Groffman et al., 2002; Addy et al., 2002). By implementing buffer restoration, many managers assume the costs of restoration will be offset by the benefits described in the scientific literature. Prevailing evidence in the form of spatial and temporal variation in buffer effectiveness suggests, however, that the water quality benefits of any buffer restoration are likely to be conditional rather than universal (e.g., Jordan et al., 1993; Hill, 1996; Correll et al., 1997; Vidon and Hill, 2004; Hefting et al., 2004). There may be a wide range of water quality benefits achieved by placing restoration activities at specific locations (e.g., Dosskey et al., 2005), but at present, there is little coordination of restoration efforts (Bernhardt et al., 2005a; Palmer et al., 2005). Given such uncertainties, it seems unlikely multiple restoration projects will necessarily provide consistent, additive water quality benefits across space or through time. This is an operating assumption yet to be evaluated across an entire watershed, however. Even so, it remains unclear whether the benefits of riparian system restoration result from nutrient interception (Lowrance et al., 1997; Weller et al., 1998), improving stream uptake potential via restoration of stream functionality (Peterson et al., 2001; Bernhardt et al., 2005b), reducing pollutant loadings by removing land from production (Dosskey, 2001), or some combination of these alternatives focused on the needs within specific landscapes. Understanding the spatial effects of these management alternatives and their potential benefits should allow greater definition of coordinated monitoring strategies and more effective prioritization of restoration spending.

Cumulative effects of economics, policy, land ownership, and ecological recovery

Complex interactions between land uses, economics, policies, and ecological processes strongly influence the timing of physical, chemical, and biological responses to conservation practices. Land use patterns reflect aggregate outcomes of rational decisions by individual landowners to optimize returns from their agricultural resources, but discrete priorities by landowners rarely result in ecologically well-integrated watersheds. Political policies affect land use change more rapidly (2 to 20 years) than the ecological processes (10 to 100 or more years) we are trying to conserve or restore. As a result, most agricultural landscapes exhibit spatial patterns of land cover and aquatic and terrestrial communities that primarily reflect the "footprint" of impermanent policies and short-term economic decisions.

Landowners, communities, and resource managers are always faced with choices of actions that sustain, deplete, or rebuild existing resources (Pitcher, 2001). Industries and societies that harvest or extract natural resources often observe gradual, long-term depletion of environmental assets. Pitcher (2001) identified three major tendencies of fisheries harvest that tend to cause

a "ratcheting effect" leading to resource depletion. The first depletion effect, which he termed "Odum's ratchet," is the tendency for past ecological conditions to become harder to restore when species (or genotypes) become extinct. As we lose biological components, ecological functions are more likely to be irreversibly changed.

The second depletion effect, termed "Pauly's ratchet," is based on the tendency for each of us to relate changes in our ecosystems to what those systems were like when we began our careers. "Accounts of former great abundance are discounted as anecdotal, methodologically naive, or are simply forgotten" [Pauly (1995), as quoted in Pitcher 2001].

The third depletion effect, termed "Ludwig's ratchet," is the tendency to increase harvest capacity through financial investment that requires continued amounts of declining resources to be harvested, generating further investment in technological capacity to harvest more resources.

Agricultural parallels are obvious, such as increased crop productivity leading to soil, water, and nutrient depletion, which requires loans for more specialized equipment and agrochemicals, which requires sustained production to repay loans required for their purchase, resulting in increased harvests from systems where soil and water resources are already becoming increasingly limited. Just as ocean fisheries have witnessed serial depletions within and among species caused by overharvest as a consequence of technological advancements in the fisheries industry, agriculture has experienced shifts in crop types and land uses as agronomic capacity becomes altered and required resources become scarce (Potter, 1998; Cochrane, 2003).

In light of the dual nature of conservation and restoration, an additional ratchet effect—"the restoration ratchet," can be added to those defined by Pitcher. This ratchet mechanism reflects the tendency to view conservation and restoration as immediately and fully effective, thereby offsetting choices leading to more intensive land use, further depleting remaining resources. In reality, the outcome of restoration may not be realized for decades after it is first implemented, and the success of conservation of existing resources remains largely unproven. Both resource managers and communities frequently make decisions based on an inherent assumption that their efforts to restore depleted resources immediately counterbalance the ongoing and often accelerating practices that depleted resources. The failure to explicitly identify realistic expectations for the timing of restoration responses inevitably leads to continued decline of natural resources and ecosystem structure and function.

Achieving greater conservation effectiveness at landscape or watershed scales

Timeframes for responses to restoration actions

Realistic timeframes for responses to ecological restoration in agricultural landscapes can be rapid (1 to 5 years), relatively fast (5 to 20 years), intermediate (20 to 50 years), slow (50 to 100 years), or extremely slow (greater than 100 years). Why do ecological processes and ecosystem components exhibit such widely differing rates of responses to restoration efforts? Many factors contribute to the timing of responses of different landscape structures, populations, and communities. Agricultural landscapes contain complex physical landforms, chemical environments, biotic communities, human communities, and histories of change. The characteristics of all of those fundamental features of agricultural land vary enormously from location to location. Therefore, it is impossible to identify exact timeframes for ecological responses to restoration efforts. We summarize several factors that shaped the responses observed in the examples we presented in Table 2.

The landscape and its physical processes set limits on potential rates of recovery in terrestrial and aquatic systems. For example, many river channels throughout the United States have been simplified and straightened. Restoration of river channels requires reconnecting historical side channels and floodplains, reestablishing channels where they have been eliminated, and restoring natural flow regimes to the extent possible. The rate of recovery of those channels will depend upon the occurrence of natural flood processes that shape and maintain river channels and their floodplains. Timing of such restorative floods will depend upon the chances of their occurrence and future weather patterns.

Rates of ecological recovery also depend upon the degree to which the system has been altered. Obviously, a slightly altered system is likely to recover much more rapidly than a landscape that has been greatly changed. For example, a farmland with large patches of native forests and relatively well connected riparian forests will respond rapidly to restoration efforts that reconnect the fragmented pieces. In contrast, a watershed that is almost completely converted to cropland, with little or no remnant native forests, will require 50 to 100 years or more to begin to support native terrestrial and aquatic communities endemic to native forests.

Recovery of ecosystems depends upon the availability of species and the resources they require. As a result, the legacy of past systems can influence recovery. For example, old-growth forests develop diverse microbial communities and organic matter in their soils. In the decades following harvest of old-growth forests, the soils contain organic matter, microbes, seeds, and invertebrates from the old forest. After repeated harvest cycles, organic matter becomes depleted, microbial diversity declines, and invertebrate communities shift to those adapted to earlier stages of forest succession.

Legacies are also important in terms of contaminants and nutrients applied and accumulated over time in agricultural landscapes. Legacies of contaminants can cause recovery to be extremely slow. Contaminants that breakdown slowly and are strongly attached to soils and particles may reside in agricultural soils for decades after agricultural practices change. The long-term trend in the persistence of DDT is an example. DDT breaks down to other chlorinated forms of hydrocarbons in 5 to 10 years, but the other forms (DDD and DDE) commonly are found in soils, organisms, and water for 30 years or more. Some chemicals, such as heavy metals like mercury and arsenic, can bind to soils and remain in storage for centuries.

Rates of ecological processes create limits for recovery. One obvious example is riparian shade. When restoration programs plant native trees along streams to restore shade, it is obvious that seedlings will provide little shade. Several decades (20 to 50 years, depending upon species) may be required to develop full canopies. If the project goal includes restoration of amounts of

Table 2. Factors that deterimine the timeframe for responses to restoration efforts.

System attributes	Recovery period			
	1–10 years	**10–50 years**	**50–100 years**	**100–1000 years**
System complexity	Simple	Simple	Complex	Complex
Control of inputs	Simple to control	Simple to control	Difficult to control	Difficult to control
Flow paths	Rapid	Intermediate	Slow	Very long and slow
Storage of nutrients, toxics, sediments, or human additions	Low	Moderate	High	High
Reproductive rates of native biota	Rapid	Rapid	Slow	Slow
Required stages of succession	Succession not required	Early stages	Mature stages	Late stages
Legacies of native ecosystems	Abundant	Abundant	Few	Few to none
Influence of alien species	Little	Slight	Extensive	Extensive and dominan
Degree of landscape alteration	Minor	Intermediate	Major	Major and irreversible

large wood in streams, more than 50 to 150 years may be required before the streamside forests begin to deliver wood to streams.

The recovery of populations depends upon rates of birth and death. Species that reproduce rapidly and produce large numbers of offspring may recover quickly after restoration is implemented. In contrast, species that reproduce and mature slowly and produce low numbers of offspring will require much longer (decades to centuries) to recover.

If land use practices causing ecological degradation continue after restoration efforts, recovery will not occur as rapidly. The degree to which pressures are placed on the recovering resources determines the rates of recovery. For example, some restoration of riparian areas involves establishment of livestock grazing exclosures. Such exclosures may encompass complete exclusion of livestock grazing or limited seasonal use. The amounts and timing of grazing can greatly influence the rates and degree of riparian and aquatic system recovery.

Couplings between the physical landscape and biological communities take time. Floodplain restoration requires reestablishment of periodic inundation. In turn, this results in changes in sediment deposition and channel change. In response, floodplain vegetation can be altered, and the composition of plant communities shifts through time as succession occurs. In turn, future floods interact with developing floodplain forests, changing the patterns that developed previously. Such interactions can proceed for decades, and outcomes of restoration efforts will reflect these changes.

Future directions to make restoration more effective

We have explored several fundamental temporal perspectives of ecological responses to restoration and conservation practices. But the larger question is how can communities and natural resource agencies become more effective in the conservation and restoration practices applied to agricultural landscapes? We suggest six major approaches that offer substantial promise to create more effective conservation and restoration: (1) Greater consideration of producer/landowner attitudes and knowledge, (2) more effective in-field practices and planning, (3) greater emphasis on effective monitoring and assessment, leading to refinement of policies and practices, (4) adoption of landscape perspectives in planning and applying conservation practices, (5) development of conservation markets, and (6) expansion of the use of alternative future scenarios.

Producer/landowner attitudes and knowledge

Agriculturalists value the culture, environmental worth, and aesthetic characteristics of their land, but personal opinions on the values of natural amenities vary. Often, one person's wildflower is another's weed. For the most part, however, those involved in agriculture embrace a desire to improve the quality and productivity of land to be passed on to future generations (Lubchenco, 1998; Wildlife Management Institute, 2006). Management philosophies guiding contemporary agricultural land use have evolved largely on the perception that composition, diversity, and ecological relations between farmed and nonfarmed land play only a small, if any, roll in productive agricultural systems (O'Riordan, 2002; Kirschenmann, 2003; Keeney and Kemp, 2004). Agricultural ecosystems are no less complex than any other ecosystem. Variability in frequency and types of land use, diverse goals of landowners, skepticism about outside intervention in management decisions, and suspicions about regulation contribute additional layers of complexity in addressing environmental issues associated with agricultural land use.

The effectiveness of conservation programs is ultimately defined by the willingness of landowners to participate and their knowledge of conservation practices and their benefits. Long-term solutions to entwined issues, such as soil erosion, water quality, and wildlife habitat, will be achieved only when conservation policies are embraced across multiple farmsteads to the watershed level. Incorporation of landowner knowledge about local issues and production challenges, coupled with forethought directed to their expectations and limitations, will elevate interest and create opportunities to improve the level of landowner knowledge required for successfully implementing conservation practices and programs. The most proficient way to get information to farmers about the benefits of conservation is to have it delivered by a neighbor who has seen success. This can then be followed up with educational activities to improve landowners' abilities to implement conservation practices successfully.

Large-scale assessments of conservation effectiveness based on sophisticated modeling are necessary for understanding effects of and refining conservation policies. Such approaches rarely, however, furnish site-specific answers to those who have invested time, labor, and trust in adoption of conservation practices on their farm. Approaches for describing on-farm or within-watershed effects of conservation are needed to strengthen and justify program participation. Many landowners who enroll in conservation programs value the environmental benefits associated with their conservation activities and want to know how well conservation practices are working on their farms. Some landowners are willing to participate in the collection of information needed to describe effectiveness of conservation policies (Wildlife Management Institute, 2006). Programs such as the Izaak Walton League's Save Our Streams (Izaak Walton League of America, 2006), where landowners are trained in sampling and identification of aquatic insects to estimate changes in water quality brought about by adoption of conservation practices, can serve as models for involving willing landowners in monitoring conservation effectiveness. Identification of specific, regionally important species as management and monitoring priorities, addressing effects of conservation practices on multifarm or watershed scales, consideration of landowner goals/limitations, as well as finding ways for willing landowners to become part of monitoring activities will improve abilities to furnish meaningful results needed to refine the performance of agricultural conservation programs.

Innovation in farm operations and waste management systems

A key factor in conservation practice effectiveness is timely adoption. Practices that are simple, easy to implement, and fit well into the agricultural operation are those most likely to be adopted by a significant number of producers. Use of precision agriculture, nutrient management, integrated pest management, on-site wastewater treatment, improved buffer designs (e.g., carbon-source trenches for enhanced denitrification), and improved livestock nutrition to reduce nitrogen and phosphorus in manures offer potential to increase the effectiveness of future restoration efforts. More specific innovations include the following:

- Use of existing in-field conservation practices (nutrient management, integrated pest management, conservation tillage, etc.) that reduce production costs and reduce resource loss from the field.
- Targeting implementation of conservation practices by identifying critical source areas within fields or landscapes.
- Elimination of agricultural subsidies that distort costs and encourage producers to over apply agricultural chemicals and farm marginal land that would not otherwise be economically productive.
- Implement conservation programs and practices that measurably improve the environment rather than those only presumed to protect the environment.
- Evaluate and improve success of conservation programs/activities by measuring improvements in environmental quality.
- Fund only conservation programs and activities that have explicit, measurable environmental goals.

Assessment and monitoring

Given the large investments of public funds in conservation and restoration actions, any prudent society would want to determine whether its efforts are successful. But observations and assessments of conservation program performance require a commitment of effort and funds. Because so few restoration programs are monitored, little information feeds back into the policy formulation and decision-making processes. As a result, adaptive management occurs most often through sequential but disconnected correction measures or emergence of new programs. A recent review of river restoration projects found 20 percent had no defined objectives, and only 10 percent included any form of assessment or monitoring (Bernhardt et al., 2005a). Post-project assessment often focuses more on implementation (e.g., how many acres or stream miles have been treated) rather than achievements of intended

environmental goals, such as measurable reduction in agricultural chemicals entering surface waters.

One of the major reasons for the low rates of monitoring and assessment is the relative cost of restoration actions versus monitoring and assessment. Most people and agencies are well intended and want to invest as much as possible in actual restoration activities. As a result, few projects dedicate funds and effort to determine whether the projects are truly successful in meeting environmental objectives, trusting that implementation of the practices alone meets program goals.

For many projects, the timing of monitoring is poorly matched to realization of expected responses. A familiar example is planting of riparian vegetation intended to reduce soil erosion, increase bank stability, increase shade, lower stream temperature, and enhance water quality, as well as the abundance, diversity, and health of fish, wildlife, and other organisms. Typically, such projects are evaluated for two to five years after establishment to determine survival of the planted vegetation. In that two- to five-year interval, it is unlikely the plant communities could develop to a stage in which they provide the intended ecological contributions (e.g., canopy cover, food inputs, wood, channel complexity). Twenty to 50 years or more is a much more realistic time horizon for recovery of many of these ecological functions.

Nonetheless, many involved in restoration understand the long-term nature of the process. In a survey of Pacific Northwest watershed councils, Bash and Ryan (2002) noted, "Many respondents indicated that short-term project assessments might not be meaningful given the time frame needed to evaluate the outcome of restoration projects."

It is highly unlikely that funds and workforce will ever be adequate to monitor and assess a large portion of the conservation or restoration actions on agricultural land. One option to provide rigorous assessment of conservation and restoration is creation of a "monitoring bank." Various projects throughout a region, from a variety of sources, could invest in a common fund that would support scientifically rigorous assessments of the major conservation and restoration actions applied to agricultural land in the region. Sites could be randomly selected from a systematic database, with factors being measured or monitored that reflect the greatest need for information identified by a ranking process that includes priorities from all agencies contributing to the monitoring bank. Conclusions drawn from study results could be scaled appropriately to the spatial and temporal scales that reflect regional applications for the intended conservation and restoration outcomes. Such an approach would eliminate duplication of monitoring efforts and maximize results from funds allocated for monitoring and assessment of conservation and restoration practices for all agencies involved in the program.

Achieving conservation effectiveness at landscape scales

The multiscale nature of watershed processes requires a watershed approach to management, but effective management of watersheds is challenging in landscapes under multiple ownerships (Allen el al., 1997). NRCS provides technical assistance to develop comprehensive resource management systems on land that may or may not be involved with conservation-oriented management. Practices implemented within the framework of a resource management system effectively protect soil and water quantity and quality as well as associated terrestrial wildlife communities. With such practices in place, aquatic species are also likely to benefit. Sedimentation of streams causes damage to habitats of all aquatic species, but that damage can be diminished when beneficial land management practices are implemented at broad scales (Lenat, 1984) and coupled with riparian conservation practices at smaller scales (Stauffer et al., 2000). Indices of biotic integrity provide insight on the effects of these practices on aquatic fauna at both scales (Lammert and Allan, 1999; Weigel et al., 2000).

Finding collaborative ways for landowners to maintain or restore connectivity of habitats should contribute to ecological restoration across wider geographic scales. For example, use of "best development practices" to improve the trajectory of amphibian populations has showed promising results when implemented cooperatively at the town level in Vermont (Calhoun et al., 2005). Maintaining contiguous riparian zones, or buffers, of adequate width along streams and rivers has been shown to correlate highly with improvements in indices of biotic integrity for aquatic fauna in Wisconsin (Weigel, 2003).

Conservation markets for regional communities

Assessments have traditionally overlooked the economic gains that can result from adoption of conservation practices. Mitigation banking has been used most widely to provide conservation benefits while also creating economic opportunities. Pollution trading is emerging as a major economic choice in response to TMDLs and other regulatory criteria. Also, state agencies are beginning to implement conservation payments to offset consumer impacts (such as large sport-utility vehicles). All of these create opportunities for farmers to implement conservation practices that potentially increase their income.

Focus on future demands and challenges rather than past practices

All too often regional assessments of conservation and restoration focus on examination of ecological conditions assumed to be related to past and current land use. Rarely are potential consequences projected for the near future (approximately 50 years). Consequently, as problems of the past are addressed, management typically fails to anticipate future challenges. Emerging resource issues are repeatedly addressed with tools designed to repair the consequences of past land use and management practices. Immediate or short-term responses often are considered to have greater likelihood of success; they often are perceived as being more credible and defensible than accepting the risk of addressing unknown changes in policy and management that might potentially affect long-term changes in resource availability or environmental conditions. As a result, decisions tend to favor near-term choices affecting small, local areas.

A proactive, longer term tool potentially applicable to management of agricultural landscapes is assessment of alternative future scenarios. Alternative-futures analysis has been used to explore future trends in the Willamette River Basin (Baker et al., 2004), as well as the San Pedro River in Arizona and Camp Pendleton in California (Steinitz et al., 2003, 2005). These assessments of future trends provide spatial projections of alternative choices about land uses and the potential environmental, economic, and social consequences of those alternatives.

A study of future alternatives for Arizona's

San Pedro River demonstrated that availability of water will have the greatest impact on future ecological conditions in this arid region (Steinitz et al., 2005). Irrigation withdrawals were projected to have the greatest potential impact on ecological processes, but policies that encouraged population growth and relaxed constraints on development also would have major impacts on water and ecological conditions. A study of land use alternatives in the upper Midwest examined people's choices for residential development in an agricultural region, finding that a majority preferred landscapes with natural vegetation and higher ecological conditions (Nassauer et al., 2004). Though questions of rural land conversion remain, the communities clearly view an ecologically healthy landscape as a more livable environment.

Environmental changes under alternative futures can be evaluated quantitatively through simulation models or observed relationships and qualitatively through expert judgment or the Delphi approach (Hulse and Gregory, 2001; Hulse et al., 2002). Mechanisms for identifying assumptions and spatial representation of alternative future scenarios are just as important, however, as are methods for analyzing alternative futures. Three approaches have been used in recent years—stakeholder-derived, expert-derived, and model-based scenarios. Each approach has strengths and weaknesses. Stakeholder processes employ citizen stakeholder groups to define assumptions about how future land and water use will unfold. Those scenarios can be used with planning processes and models to produce maps of potential future land and water use, translating the stakeholder assumptions into mapped form. The stakeholder approach has the advantages of citizen involvement, greater political plausibility, and an increased likelihood of institutional acceptance. But stakeholder–driven processes have one disadvantage: They do not statistically quantify the likelihood of various alternatives, and the number of alternatives produced (three to ten) is typically limited.

A second common approach for creating mapped alternative futures is expert judgment, with professionals in the biophysical and social sciences defining processes and rates of transition that may determine future land and water use conditions. Alternative futures produced from expert judgment have the advantage of quantifiable statistical likelihood (based on the larger number of alternatives produced), but suffer from unclear political plausibility and a lack of citizen involvement, which often limits their credibility in affected communities.

Simulation modeling has been used to define alternative futures by representing the rules by which people make decisions and then projecting probable effects across the landscape. Simulation models can produce thousands of possible future landscapes, with the advantage of representing the statistical likelihood of various alternatives. An additional advantage of simulation models is the ability to create and run new alternatives quickly.

Trajectories of land use and environmental change from 1850 to 2050 were developed for the 30,000-square-kilometer (11,583-square-mile) Willamette River Basin in Oregon, a basin comprised of approximately 25 percent agricultural land, 65 percent forest land, 6 percent urban land, and 4 percent rural residential land (Baker et al., 2004). Human population in the basin is expected to increase from 2.2 million to more than 4 million by 2050. Three spatially explicit future scenarios were developed by a group of stakeholders: (1) a Plan Trend 2050 scenario in which current policies and practices continue through 2050, (2) a Development 2050 scenario in which market forces are allowed to influence land use change and current land use policies are relaxed, and (3) a Conservation 2050 scenario in which additional, plausible conservation and restoration practices are implemented. Scenario outcomes were evaluated on the basis of land cover change, water availability, and models of ecological conditions for fish, macroinvertebrate, and wildlife communities.

Incorporation of conservation practices in Conservation 2050 enhanced wildlife habitat without significantly altering the function of the agricultural system. Development 2050 also showed local improvement in wildlife habitat due to increases in natural vegetation associated with the developed environment. Plan Trend 2050 indicated little change in habitat quality because few modifications were made to agricultural land.

The Willamette Valley contained approximately 240,00 hectares (620,000 acres) of prime farmland in 1990, almost all of which remained in agricultural production (1 percent was converted). Under the Development 2050 scenario, approximately 25 percent of prime farmland would be converted to other uses, leading to fragmentation and conversion of agricultural fields. Under the Conservation 2050 scenario, 15 percent of prime farmland would be converted to field borders, low-input crops in sensitive areas, and conversion of cropland to native vegetation. Development scenarios tended to prefer areas of prime farmland, while restoration activities tended to focus on lower quality, less productive farmland.

One of the most important findings of the alternative-futures analysis is that both the Plan Trend 2050 and Development 2050 scenarios show either little change or continued decline in natural resources (Figure 1). In sharp contrast, indicators of natural resource condition improve substantially under the stakeholders' assumptions about plausible restoration measures in the conservation 2050 scenario, recovering 20 to 70 percent of the losses sustained since settlement in the mid-1800s. Citizens and decision-makers in the basin now have geographic projections over the next 50 years, indicating conservation and restoration practices are likely to produce significant ecosystem benefits while accommodating the projected increase in the human population.

An agent-based model, Evoland (Evolving Landscapes), was developed to examine ecological and economic consequences of alternative futures for floodplains and riparian areas of the Willamette River Basin (John Bolte, Oregon State University, personal communication). This modeling approach allows rapid analysis of many alternative futures, measurement of variance based on probabilities of land use choices, and modification of assumptions and policies defined by user groups. Results of modeling alternative policies clearly demonstrate conservation and restoration policies can be effective in restoring ecological function in the long run (20 to 40 years), but ecological conditions respond to conservation and restoration actions more slowly than they do in response to economic and social policies. An additional concern raised focuses on effectiveness of adaptive management. If policies were implemented that would result in short-term economic gain, but cause floodplain and riparian degradation not evident for 20 to 30 years, adaptive management would be ineffective in the face of the substantial financial investments that would have occurred before the undesired outcomes

were realized. The timing of restoration outcomes will be constrained by the competing processes of intensified land use and land use conversion.

Making decisions for generations

In his 1999 book *The Clock of the Long Now*, Stewart Brand addresses the challenge of incorporating different time scales into the decision-making process. He asks, "How do we make long-term thinking automatic and common instead of difficult and rare, and how do we make the taking of long-term responsibility inevitable?" Tools and ways of thinking have to be changed so that the "long now" is inherent in the management questions asked and the solutions explored. Because environmental and social consequences of modern agricultural production reach from the heart of this continent into coastal and marine ecosystems, we can no longer measure agricultural accomplishments simply on economic returns brought about by traditional farm products.

Unfortunately, there has been an inclination to define the agricultural landscape as being composed of either "working" or "conservation" land. This regrettable distinction, born in part by the structure of existing conservation programs, fails to recognize that all land, regardless of production status, is part of the "working" agricultural ecosystem. An economically viable, environmentally sound, and, therefore, socially supportable agricultural industry will be possible only when agriculture protects and even perhaps enhances the natural and cultural resources upon which it stands.

Budgetary constraints increasingly force decisions affecting how conservation programs are designed and administered. Successful conservation policies can be publicly and politically supported only when their effectiveness is known. To gain such knowledge requires an unrelenting commitment to calculate both immediate as well as long-term effectiveness of programs and refine conservation policies as information becomes available. The reality that must be faced is hard numbers that either support or disprove the success of conservation and restoration activities will not appear quickly, nor will they come without a commitment to fund the research required to define acceptable solutions.

Traditional conservation and restoration practices will continue to be used. Well-intended land-

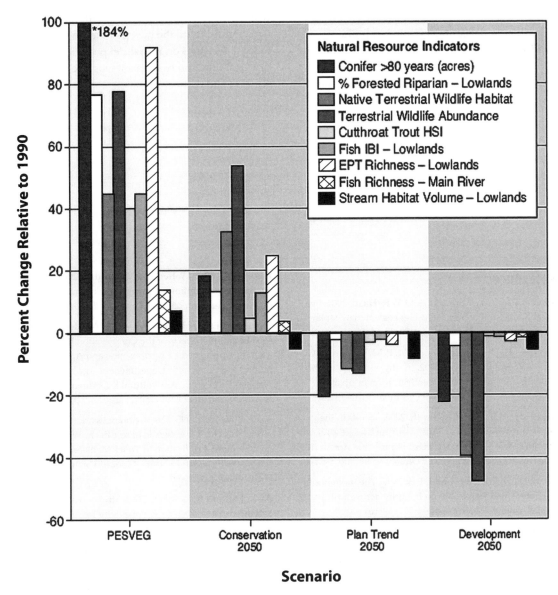

Figure 1. Percent change in measures of natural resource condition in the three future (Conservation, Plan Trend, Development) and pre EuroAmerican (PESVEG) scenarios relative to 1990 land use and cover in the Willamette River Basin. Source: Baker et al., 2004.

owners and community groups will continue to try to sustain and restore declining resources in the face of growing human populations and their need for agricultural commodities. There will be no easy answers, and good intentions alone will not suffice. The tremendous power of the ratchet effects in place in society—extinction of species, generational views of resource abundance and landscape condition, and economic pressures that require continued or accelerated commodity production—must be faced. And through the temporal perspectives explored in this paper, the timeframes of both conservation and restoration must be carefully and clearly explained to avoid the ratchet effect of assuming these practices will be immediately and fully effective. It cannot be assumed that continued or accelerated demands on natural resources can be counterbalanced by conservation and restoration measures alone. The uncertainty in that assumption, even when balanced with more realistic expectations of timeframes, must be adjusted with a "margin of safety" for natural resources, just as engineers would use in designing any road or building. Landowners and resource managers must balance the immediate impacts of their actions against the current rates of resource restoration. The actions taken today determine the extent to which the

world will sustain the next generation. We have a responsibility to the next generation to base today's decisions on realistic expectations about practical timeframes for achieving ecological restoration and conservation.

References

Adams, S.B., M.L. Warren Jr., and W.R. Haag. 2004. Spatial and temporal patterns in upper coastal plain streams, Mississippi, USA. *Hydrobiologia* 528: 45-61.

Addy, K., D. Q. Kellogg, A. J. Gold, P. M. Groffman, G. Ferrendo, and C. Sawyer. 2002. In situ push-pull method to determine ground water denitrification in riparian zones. *Journal of Environmental Quality* 31:1,017-1,024.

Alexander, R.B., and R.A. Smith. 2006. Trends in the nutrient enrichment of U.S rivers during the late 20th century and their relation to changes in probable stream trophic conditions. *Limnology and Oceanography* 51:639-654.

Allan, D., D. Erickson, and J. Fay. 1997. The influence of catchment land use on stream integrity across multiple spatial scales. *Freshwater Biology* 37: 149-161.

Allen, A.W., and M.W. Vandever. 2003. A national survey of Conservation Reserve Program (CRP) participants on environmental effects, wildlife issues, and vegetation management on program lands. Biological Science Report, USGS/BRD/BSR—2003-0001. U.S. Government Printing Office, Denver, Colorado. 51 pp.

Allen, A.W., and M.W. Vandever, editors. 2005. The Conservation Reserve Program—planting for the future. Proceedings of a National Conference, Fort Collins, Colorado, June 6-9, 2004. Scientific Investigations Report 2005-5145. U.S. Geological Survey, Fort Collins, Colorado. 248 pp.

Baker, J.P., D.W. Hulse, S.V. Gregory, D. White, J. Van Sickle, P.A. Berger, D. Dole, and N.H. Schumaker. 2004. Alternative futures for the Willamette river basin. *Ecological Applications* 14:313–324.

Baker, M.E., D.E. Weller, and T.E. Jordan. 2006. Improved methods for quantifying potential nutrient interception by riparian buffers. *Landscape Ecology* 21:1,327– 1,345.

Baker, M.E., D.E. Weller, and T.E. Jordan. 2007. Effects of stream map resolution on patterns of riparian buffers and nutrient retention potential. *Landscape Ecology* (in press).

Bangsund, D.A., F.L. Leistritz, and N.M. Hodur. 2002. Rural economic effects of the Conservation Reserve Program in North Dakota. Agribusiness and Applied Economics Report No 497. Department of Agribusiness and Applied Economics, Agricultural Experiment Station, North Dakota State University. Fargo. 105 pp.

Bartholow, J.M. 1995. The stream network temperature model (SNTEMP): A decade of results. In Workshop on Computer Application in Water Management. Water Resources Research Institute, Colorado State University, Fort Collins. pp. 57–60.

Bash, J.S., and C.M. Ryan. 2002. Stream restoration and enhancement projects: is anyone monitoring? *Environmental Management* 29(6): 877–885.

Bernhardt, E. S., M.A. Palmer, J.D. Allan, G.Alexander, K. Barnas, S. Brooks, J. Carr, S. Clayton, C. Dahm, J. Follstad-Shah, D. Galat, S. Gloss, P. Goodwin, D. Hart, B. Hassett, R. Jenkinson, S.Katz, G.M.Kondolf, P. S. Lake, R. Lave, J. L.Meyer, T.K. O'Don. 2005a. Synthesizing U.S. river restoration efforts. *Science* 308: 636-637.

Bernhardt, E.S., GE. Likens, R.O. Hall, D.C. Buso, S.G. Fisher, T.M. Burton, J.L. Meyer, W.H. McDowell, M.S.Mayer, W.B. Bowden, S.G. Findlay, K.H. MacNeale, R.S. Telzer, and W.H. Lowe. 2005b. Can't see the forest for the stream? In-stream processing and terrestrial nitrogen exports. *Bioscience* 55: 219-230.

Blann, K.L., J.F. Nerbonne, and B. Vondracek. 2002. Relationship of riparian buffer type to physical habitat and stream temperature. *North American Journal of Fisheries Management* 22:441-451.

Boesch, D.F., R.B. Brinsfield, and R.E. Magnien. 2001. Chesapeake Bay eutrophication: Scientific understanding, ecosystem restoration, and challenges for agriculture. *Journal of Environmental Quality* 30: 303-320.

Bohlke, J.K., and J.M. Denver. 1995. Combined use of groundwater dating, chemical, and isotopic analyses to resolve the history and fate of nitrate contamination in two agricultural watersheds, Atlantic coastal plain, Maryland. *Water Resources Research* 31: 2,319-2,339.

Bohlke, J.K., J.W. Harvey, and M.A. Voytek. 2004. Reach-scale isotope tracer experiment to quantify denitrification and related processes in a nitrate-rich stream, mid-continent United States. *Limnology and Oceanography* 49: 821-838.

Bond, N.R., and P.S. Lake. 2003. Local habitat restoration in streams: Constraints on the effectiveness of restoration for stream biota. *Ecological Management and Restoration* 4: 193-198.

Borman, M.M., and Larson, L.L. 2003. A case study of river temperature response to agricultural land use and environmental thermal patterns. *Journal of Soil and Water Conservation* 58: 8-12.

Brand, S. 1999. The clock of the long now: Time and responsibility. Basic Books, New York, New York.

Burton G.A. Jr., D. Gunnison, and G.R. Lanza. 1987. Survival of pathogenic bacteria in various freshwater sediments. *Applied Environmental Microbiology* 53: 633-638.

Cade, B.S., M.W. Vandever, A.W. Allen, and J.W. Terrell. 2005. Vegetation changes over 12 years in ungrazed and grazed Conservation Reserve Program grasslands in the Central and Southern Great Plains. In A.W. Allen and M.W. Vandever, editors, The Conservation Reserve Program – Planting for the Future: Proceedings of a National Conference, Fort Collins, CO, June 6-9 2004. Scientific Investigations Report 2005-5145. U.S. Geological Survey, Fort Collins, Colorado. pp. 106-119.

Calamari, D., E. Zuccato, S. Castiglioni, R. Bagnati, and R. Fnelli. 2003. Strategic survey of therapeutic drugs in the Rivers Po and Lambro in Northern Italy. *Environmental Science and Technology* 37: 1,241-1,248.

Calhoun, A.J.K, N.A. Miller, and M.W. Klemens. 2005. Conserving pool-breeding amphibians in human-dominated landscapes through local implementation of Best Development Practices. *Wetlands Ecology and Management* 13(3): 291-304.

Cessna, A.J., and J.A. Elliott. 2004. Seasonal variation of herbicide concentrations in prairie farm dugouts. *Journal of Environmental Quality* 33:302-315.

Cirmo C.P., and J.J. McDonnell. 1997. Linking the hydrologic and biogeochemical controls of nitrogen transport in near-stream zones of temperate forested catchments: a review. *Journal of Hydrology* 199: 88-120.

Cochrane, W.W. 2003. The curse of American agricultural abundance: A sustainable solution. University of Nebraska Press, Lincoln. 154 pp.

Collins, R., S. Elliott, and R. Adams. 2005. Overland flow delivery of faecal bacteria to a headwater pastoral stream. *Journal of Applied Microbiology* 99:126-132.

Colvin, R. 2005. Fish and amphibian use of seasonal drainages within the upper Willamette River Basin, Oregon. MS thesis, Oregon State University, Corvallis. 131 pp.

Correll, D.L., T.E. Jordan, and D.E. Weller. 1997. Failure of agricultural riparian buffers to protect surface waters from groundwater nitrate contamination. In J. Gibert, J. Mathieu, and F. Fournier, editors, Groundwater/Surface Water Ecotones: Biological and Hydrological Interactions and Management Options. Cambridge University Press, Cambridge, United Kingdom. pp. 162-165.

Costanza, R., R. d'Arge, R. de Groots, S. Farber, M. Grasso, B. Hannon, K. Limburg, S. Naeem, R.V. O'Neill, J. Paruelo, R.T. Raskins, P. Sutton, and M. van den Belt. 1997. The value of the world's ecosystem services and natural capital. *Nature* 387: 253-260.

Costanza, R., H. Daly, C. Folke, P. Hawken, C.S. Holling, A.J.D. McMichael, D. Pimentel, and D. Rapport. 2000. Managing our environmental portfolio. *Bioscience* 50(2): 149-155.

Diaz-Cruz, M.S., M.J. Lopez de Alda, and D. Barcelo. 2003. Environmental behavior and analysis of veterinary and human drugs in soils, sediments and sludge. *Analytical Chemistry* 22: 340-351.

Dillaha, T.A., R.B. Reneau, S. Mostaghimi, and D. Lee. 1989. Vegetative filter strips for agricultural nonpoint source pollution control. *Transactions, American Society of Agricultural Engineers* 32(2): 491 496.

Dosskey, M.G. 2001. Toward quantifying water pollution abatement in response to installing buffers on crop land. *Environmental Management* 28: 577-598.

Dosskey, M.G., D.E. Eisenhauer, and M.J. Helmers. 2005. Establishing conservation buffers using precision information. *Journal of Soil and Water Conservation* 60(6): 349-354.

Ebersole, J.L., W.J. Liss, and C.A. Frissel. 2003. Coldwater patches in warm streams: Physiochemical characteristics and the influence of shading. *Journal of the American Water Resources Association* 39(2): 355–368.

Elliott, J.A, A.J. Cessna, and C.R. Hilliard. 2001. Influence of tillage system on water quality and quantity in prairie pothole wetlands. *Canadian Water Resources Journal* 26: 165-181.

Feather, P., D. Hellerstein, and L. Hansen. 1999. Economic valuation of environmental benefits and the targeting of conservation programs. Agricultural Economic Report No. 778. Resource Economics Division, Economic Research Service, U.S. Department of Agriculture, Washington, D.C. 56 pp.

Fitzpatrick, F.A., B.C. Scudder, B.N. Lenz, and D.J. Sullivan. 2001. Effects of multi-scale environmental characteristics on agricultural stream biota in eastern Wisconsin. *Journal of the American Water Resources Association* 37: 1,489-1,508.

Focazio, M.J., L.N. Plummer, J.K. Bohlke, E. Busenberg, L.J. Bachman, and D.S. Powars. 1997. Preliminary estimates of residence times and apparent ages of ground water in the Chesapeake Bay watershed and water-quality data from a survey of springs. Water-Resources Investigations Report 97-4225. U.S. Geological Survey, Reston, Virginia.

Frenkel, R.E., and J.C. Morlan. 1991. Can we restore our salt marshes? Lessons from the Salmon River, Oregon. *Northwest Environmental Journal* 7: 119–135.

Fritcher, S.C., M.A. Rumble, and L.D. Flake. 2004. Grassland bird densities in seral stages of mixed grass prairie. *Journal of Range Management* 57(4): 351-357.

Glozier, N.E., J.A. Elliott, B.Holliday, J. Yarotski and B. Harker. 2006. Water quality trends and characteristics in a small agricultural watershed: South Tobacco Creek, Manitoba 1992-2001. Environment Canada, Ottawa, Ontario.

General Accounting Office. 1995. Agriculture and the environment, information on the characteristics of selected watershed projects. Report to the U.S. Senate Committee on Agriculture, Nutrition, and Forestry. Washington, D.C. 65 pp.

Groffman, P.M., A.J. Gold, D.Q. Kellogg, and K. Addy. 2002. Mechanisms, rates and assessment of N2O in groundwater, riparian zones and rivers. In J. van Ham, A. P. M. Baede, R. Guicherit, J.G.F.M. Williams-Jacobse, editors, *Non-CO2 Greenhouse Gases: Scientific Understanding, Control Options and Policy Aspects*, Millpress, Rotterdam, The Netherlands. pp. 159-166.

Halling-Sorensen, B., S. Nors Nielsen, P.F. Lanzky, F. Ingerslev, H.C. Holten Lutzheft, and S.E. Jorgensen. 1998. Occurance, fate and effects of pharmaceutical substances in the environment--a review. *Chemosphere* 36: 357-393.

Halvorson, A.D., and A.L.Black. 1985. Long term dryland crop responses to residual phosphorus fertilizer. *Soil Science Society of America Journal* 49: 928-933.

Harding, J.S., E.F. Benfield, P.V. Bolstad, G.S. Helfman, and E.B.D. Jones III. 1998. Stream biodiversity: The ghost of land use past. *Proceedings, National Academy of Sciences,USA* 95: 14,843–14,847.

Hart, D.D., T.E. Johnson, K.L. Bushaw-Newton, R.J. Horwitz, A.T. Bednarek, D.F. Charles, D.A. Kreeger, and D.J. Velinsky. 2002. Dam removal: Challenges and opportunities for ecological research and river restoration. *BioScience* 52: 669–682.

Hassett, B., M.A. Palmer, E.S. Bernhardt, S. Smith, J. Carr, and D.D. Hart. 2005. Restoring watersheds project by project: Trends in Chesapeake Bay tributary restoration. *Frontiers in Ecology & the Environment* 3(5): 259-267.

Haufler, J.B., editor. 2005. Fish and wildlife benefits of Farm Bill conservation programs: 2000-2005 update. Technical Review 05-2. The Wildlife Society, Bethesda, Maryland. 205 pp.

Hefting, M., J.C. Clement, D. Dowrick, A.C. Cosandey, S.Bernal, C. Cimpian, A. Tartur, T.P. Burt, and G. Pinay. 2004. Water table elevation controls on soil nitrogen cycling in riparian wetlands along a European climatic gradient. *Biogeochemisty* 67: 113-134.

Henning, J.A. 2005. Floodplain emergent wetlands as rearing habitat for fishes and the implications for wetland enhancement. M.S. thesis, Oregon State University, Corvallis. 40 pp.

Hill, A.R. 1996. Nitrate removal in stream riparian zones. *Journal of Environmental Quality* 25: 743-755.

Holdren, G.C. Jr., and D.E. Armstrong. 1980. Factors affecting phosphorus release from intact lake sediment cores. *Journal of the American Chemical Society* 14: 79-85.

Hulse, D., J. Eilers, K, Freemark, D. White, and C. Hummon. 2000. Planning alternative future landscapes in Oregon: evaluating effects on water quality and biodiversity. *Landscape Journal* 19(2): 1-19.

Hulse, D.H., and S.V. Gregory. 2001. Alternative futures as an integrative framework for riparian restoration of large rivers. In V.H. Dale And R. Haeuber, editors, Applying Ecological Principles To Land Management. Springer-Verlag, New York, New York. pp. 194-212.

Hulse, D.H., S.V. Gregory, and J. Baker, editors. 2002. Willamette River Basin planning atlas: Trajectories of environmental and ecological change. Oregon State University Press, Corvallis. 178 pp.

Independent Multidisciplinary Science Team. 2004. Oregon's water temperature standard and its application: causes, consequences, and controversies associated with stream temperature. Technical Report 2004-1 to the Oregon Plan for Salmon and Watersheds. Oregon Watershed Enhancement Board, Salem.

Inoue, M., and S. Nakano. 2001. Fish abundance and habitat relationships in forest and grassland streams, northern Hokkaido, Japan. *Ecological Research* 16(2): 233-247.

Izaak Walton League of America. 2006. Watershed programs. http://www.iwla.org/index.php

Jensen, H.S., and F.O. Andersen. 1992. Importance of temperature, nitrate and pH for phosphorus released from aerobic sediments of four shallow eutrophic lakes. *Liminology and Oceanography* 37: 577-589.

Johnson P.A. 2002. Incorporating road crossings into stream and river restoration projects. *Ecological Restoration* 20: 270-277.

Johnson, J., and B. Maxwell. 2001. The role of the Conservation Reserve Program in controlling rural residential development. *Journal of Rural Studies* 17: 323-332.

Jordan, T.E., D.L. Correll, and D.E. Weller. 1993. Nutrient interception by a riparian forest receiving cropland runoff. *Journal of Environmental Quality* 22: 467-473.

Kamler, J.F., W.B. Ballard, and D.A Swepston. 2001. Range expansion of mule deer in the Texas panhandle. *The Southwestern Naturalist* 46(3): 378-379.

Kamler, J.F., W.B. Ballard, E.B. Fish, P.R. Lemons, K. Mote, and C.C. Perchellet. 2003. Habitat use, home ranges, and survival of swift foxes in a fragmented landscape: conservation implications. *Journal of Mammalogy* 84(3): 989-995.

Kanehl, P.D., J. Lyons, and J.E. Nelson. 1997. Changes in the habitat and fish community of the Milwaukee River, Wisconsin, following removal of the woolen mills dam. *North American Journal of Fisheries Management* 17: 387-400.

Keeney, D, and L. Kemp. 2004. A new agricultural policy for the United States. In S.S. Light, editor, The Role of Biodiversity Conservation in the Transition to Rural Sustainability. IOS Press, Washington, D.C. pp. 29-47.

Kirschenmann, F. 2003. The current state of agriculture: Does it have a future? In N. Wirzba, editor, *The Essential Agrarian Reader: The Future of Culture, Community and the Land*. University Press of Kentucky, Lexington. pp. 101-120.

Kolpin, D.W., E.T. Furlong, M.T. Meyer, M.T., E.M. Thurman, S.D. Zaugg, L.B. Barber, and H.T. Buxton. 2002. Pharmaceuticals, hormones, and other organic wastewater contaminants in U.S. streams, 1999-2000: A national reconnaissance. *Environmental Science and Technology* 36: 1,202-1,211.

Krapu, G.L., D.A. Brandt, and R.R. Cox, Jr. 2004. Less waste corn, more land in soybeans and the switch to genetically modified crops: Trends with important implications for wildlife management. *Wildlife Society Bulletin* 32(1): 127-136.

LaLiberte, P., and D.J. Grimes. 1982. Survival of Escherichia coli in lake bottom sediment. *Applied Environmental Microbiology* 43: 623-628.

Lambert, D., P. Sullivan, R. Claassen, and L. Foreman. 2006. Conservation-compatible practices and programs: Who participates? Economic Research Report No. 14. U.S. Department of Agriculture, Economic Research Service, Washington, D.C. 43 pp.

Lammert, M., and J. D. Allan. 1999. Assessing biotic integrity of streams: Effects of scale in measuring the influence of land us/cover and habitat structure of fish and macroinvertebrates. *Environmental Management* 23: 257-270.

Lenat, D.R. 1984. Agriculture and stream water quality: a biological evaluation of erosion control practices. *Environmental Management* 8: 333–344.

Lewis, T.E., D.W. Lamphear, D.R. McCanne, A.S. Webb, J.P. Krieter, and W.D. Conroy. 2000. Regional assessment of stream temperatures across northern California and their relationship to various landscape-level and site-specific attributes. Forest Science Project. Humboldt State University Foundation, Arcata, California.

Li, H.W., T.N. Pearsons, C.K. Tait, J.L. Li, and J.C. Buckhouse. 1994. Cumulative effects of riparian disturbances along high desert trout streams of the John Day Basin, Oregon. *Transactions of the American Fisheries Society* 123: 627–640.

Lindsay, M.E., M. Meyer, and E.M. Thurman. 2001. Analysis of trace levels of sulphonamide and tetracycline antimicrobials in groundwater and surface water using solid-phase extraction and liquid chromatography/mass spectrometry. *Analytical Chemistry* 73: 4,640-4,646.

Lindsey, B.D., S.W. Phillips, C.A. Donnelly, G.K. Speiran, L.N. Plummer, J.K. Bohlke, M.J. Focazio, W.C. Burton, E. Busenberg. 2003. Residence times and nitrate transport in groundwater discharging to streams in the Chesapeake Bay Watershed. Water-Resources Investigations Report 03-4035. U.S. Geological Survey, Reston, Virginia.

Loomis, J., P. Kent, L. Strange, K. Fausch, and A. Covich. 2000. Measuring the total economic value of restoring ecosystem services in an impaired river basin: Results from a contingent valuation survey. *Ecological Economics* 33: 103-117.

Lowrance R.R., L.S. Altier, J.D. Newbold, R.R. Schnabel, P.M Groffman, J.M. Denver, D.L. Correll, J.W. Gilliam, J.L. Robinson, R.B. Brinsfield, K.W. Staver, W. Lucas, and A.H. Todd. 1997. Water quality functions of riparian forest buffers in Chesapeake Bay watersheds. *Environmental Management* 21: 687-712.

Lubchenco, J. 1998. Entering the century of the environment: a new social contract for science. *Science* 279(23): 491-497.

Maloney, S.B., A.R. Tiedemann, D.A. Higgins, T.M. Quigley, and D.B. Marx. 1999. Influence of stream temperature characteristics and grazing intensity on stream temperatures in eastern Oregon. General Technical Report PNW-GTR-459. U.S. Department of Agriculture, Forest Service, Pacific Northwest Research Station, Portland, Oregon.

Marsden, M.W. 1989. Lake restoration by reducing external phosphorus loading: The influence of sediment phosphorus release. *Freshwater Biology* 29: 131-162.

Marsh, N., C.J. Rutherford, and S. Bunn. 2005. The role of riparian vegetation in controlling stream temperature in a southeast Queensland stream. Technical Report 05/3. Cooperative Research Centre for Catchment Hydrology, Victoria, Australia. 22 pp.

Maulé, C.P., and J.A. Elliott. 2005. Effect of hog manure injection upon soil productivity and water quality; Part I, Perdue site, 1999-2004. ADF Project 98000094. Saskatchewan Agriculture Development Fund, Regina.

McCoy, T.D., M. R. Ryan, L.W. Burger, and E.W. Kurzejeski. 2001. Grassland bird conservation: CP1 vs. CP2 plantings in Conservation Reserve Program fields in Missouri. *American Midland Naturalist* 145(1): 1-17.

McIssac, G.F., M.B. David, G.Z. Gertner, and D.A. Goolsby. 2001. Eutrophication: nitrate flux in the Mississippi River. *Nature* 414: 166-167.

McMichael, G. A., J.A. Vucelick, C.S. Abernethy, and D.A. Neitzel. 2004. Comparing fish screen performance to physical design criteria. *Fisheries* 29(7): 10-16

Michel, R.L. 1992. Residence times in river basins as determined by analysis of long term tritium records. *Journal of Hydrology* 130: 367-378.

Minns, C. K., J.R. M. Kelso, and R.G. Randall. 1996. Detecting the response of fish to habitat alterations in freshwater ecosytems. *Canadian Journal of Fisheries and Aquatic Sciences* 53: 403-414

Molenat, J., and C. Gascuel-Odoux. 2002. Modelling flow and nitrate transport in groundwater for the prediction of water travel times and of consequences of land use evolution on water quality. *Hydrological Process* 16: 479-492.

Mosley, M.P. 1983. Variability of water temperatures in the braided Ashley and Rakaia rivers. *New Zealand Journal of Marine and Freshwater Research* 17: 331–342.

Muirhead, R.W., R.P. Collins, and P.J. Bremer. 2006 Numbers and transported state of Escherichia coli in runoff direct from fresh cowpats under simulated rainfall. *Letters in Applied Microbiology* 42: 83-88

Nagasaka, A., and F. Nakamura. 1999. The influences of land use changes on hydrology and riparian environment in a northern Japanese landscape. *Landscape Ecology* 14(6): 543-556.

Nassauer, J.I., J.D. Allan, T. Johengen, S.E. Kosek, and D. Infante. 2004. Exurban residential subdivision development: Effects on water quality and public perception. *Urban Ecosystems* 7(3): 267-281.

National Research Council. 2002. Riparian areas: Functions and strategies for management. National Academy Press, Washington, D.C.

Oenema, O., and C.W.J. Roest. 1998. Nitrogen and phosphorus losses from agriculture into surface waters; the effects of policies and measures in the Netherlands. *Water Science Technology* 37(2): 19-30.

O'Riordan, T. 2002. Protecting beyond the protected. In T. Riordan and S. Stoll-Kleemann, editors, Biodiversity, Sustainability and Human Communities: Protecting Beyond the Protected. Cambridge University Press, Cambridge, United Kingdom. pp. 3-29.

Palmer, M.A., E.S. Bernhardt, J.D. Allan, P.S. Lake, G. Alexander, S. Brooks, J. Carr, S. Clayton, C.N. Dahm, J. Follstad-Shah, D.L. Galat, S.G. Loss, P. Goodwin, D.D. Hart, B. Hassett, R. Jenkinson, G.M. Kondolf, R. Lave, J.L. Meyer, T.K. O'Donnell, L. Pagano and E. Sudduth. 2005. Standards for ecologically successful river restoration. *Journal of Applied Ecology* 42: 208-217.

Parkyn, S.M., R.J. Davies-Colley, N.J. Halliday, K.J. Costley, and G.F. Croker. 2003. Planted riparian buffer zones in New Zealand: Do they live up to expectations? *Restoration Ecology* 11(4): 436-447.

Pauly, D., T.J. Pitcher, and D. Preikshot, editors. (1998). Back to the future: Reconstructing the Strait of Georgia ecosystem. *UBC Fisheries Centre Research Reports* 6(5):100.

Pess, G. R., M.E. McHugh, D. Fagen, P. Stevenson, and J. Drotts. 1998. Stillaguamish salmonid barrier evaluation and elimination project—phase III. Final report to the Tulalip Tribes, Marysville, Washington.

Peterjohn, W.T., and D.L. Correll. 1984. Nutrient dynamics in an agricultural watershed: observations on the role of a riparian forest. *Ecology* 65: 1,466-1,475.

Peterson, B.J., W.M. Wolheim, P.J. Mulholland, J.R. Webster, J.L. Meyer, J.L. Tank, E. Marti, W.B. Bowden, H.M. Valett, A.E. Hershey, W.H. McDowell, W.K. Dodds, S.K. Hamilton, S. Gregory, and D.D. Morrall. 2001. Control of nitrogen export from watersheds by headwater streams. *Science* 292: 86-90.

Petranka, J.W., S.S. Murray, and C.A. Kennedy. 2003. Responses of amphibians to restoration of a southern Appalachian wetland: perturbations confound post-restoration assessment. *Wetlands* 23: 278–290.

Pitcher T.J. 2001. Fisheries managed to rebuild ecosystems: reconstructing the past to salvage the future. *Ecological Applications* 11(2): 601-617.

Pinay, G., J.C. Clement, and R.J. Naiman. 2002. Basic principles and ecological consequences of changing water regimes on nitrogen cycling in fluvial systems. *Environmental Management* 30: 481-491.

Poole, G.C., and C. Berman. 2001. An ecological perspective on in-stream temperature: natural heat dynamics and mechanisms of human-caused thermal degradation. *Environmental Management* 27(6): 787–802.

Potter, C. 1998. Against the grain: Agri-environmental reform in the United States and the European Union. CAB International, Wallinford, United Kingdom. 194 pp.

Quinn, J.M., R.B. Williamson, R.K. Smith, and M.L. Vickers. 1992. Effects of riparian grazing and channelizsation on streams in Southland, New Zealand. 2. Benthic invertebrates. *New Zealand Journal of Marine and Freshwater Research* 26: 259-273.

Read, D.W.L., E.D. Spratt, L.D. Bailey, E.G. Warder, and W.S. Ferguson. 1973. Residual value of phosphatic fertilizer on Chernozemic soils. *Canadian Journal of Soil Science* 53: 389- 398.

Rodgers, R.D. 1999. Why haven't pheasant populations in western Kansas increased with CRP? *Wildlife Society Bulletin* 27(3): 654-665.

Roni, P., and T. P. Quinn. 2001. Density, and size of juvenile salmonids in response to placement of large woody debris in western Oregon, and Washington streams. *Canadian Journal of Fisheries and Aquatic Sciences* 58: 282–292.

Sarr, D.A. 2002. Riparian livestock exclosure research in the western United States: a critique and some recommendations. *Environmental Management* 30(4): 516–526.

Saunders, W.C. 2006. Improved grazing management increases terrestrial invertebrate inputs that feed trout in Wyoming rangeland streams. Master's thesis, Colorado State University, Fort Collins.

Saunders, W.C., and K.D. Fausch. 2006. A field evaluation of the effects of improved grazing management on terrestrial invertebrate inputs that feed trout in Wyoming rangeland streams. Final report. Natural Resources Conservation Service, Portland, Oregon.

Schilling, K.E., and J. Spooner. 2006. Effects of watershed-scale land use change on stream nitrate concentrations. *Journal of Environmental Quality* 35:2,132–2,145.

Schippers, P., H. van de Weerd, J. de Klein, B. de Jong and M. Scheffer. 2006. Impacts of agricultural phosphorus use in catchments on shallow lake water quality: About buffers, time delays and equilibria. *Science of the Total Environment* 369: 280-294.

Sheppard, S.C., M.I. Sheppard, J. Long, B. Sanipelli, and J. Tait. 2006. Runoff phosphorus retention in vegetated field margins on flat landscapes. *Canadian Journal of Soil Science* (in press).

Shields, F.D. Jr., S.S. Knight, and J.M. Strofleth. 2006. Large wood additions for aquataic habitat rehabilitation in an incised, sand-bed stream, Little Topashaw Creek, Mississippi. *River Research and Applications* (in press).

Soil and Water Conservation Society. 2006. Final report from the Blue Ribbon Panel conducting an external review of the U.S. Department of Agriculture's Conservation Effects Assessment Project. Ankeny, Iowa. 25 pp.

Stalnacke, P.A., Grimvall, C. Libiseller, M. Laznik, and I. Kokorite. 2003. Trends in nutrient concentrations in Latvian rivers and the response to dramatic change in agriculture. *Journal of Hydrology* 283: 184-205.

Stammler, K.L. 2005. Agricultural drains as fish habitat in southwestern Ontario. Master's thesis. University of Guelph, Guelph, Ontario. 45 pp.

Stanford, J.A., J.V. Ward, W.J. Liss, C.A. Frissell, R.N. Williams, J.A. Lichatowich, and C.C. Coutant. 1996. A general protocol for restoration of regulated rivers. *Regulated Rivers* 12: 391–413.

Stauffer, J.C., R.M. Goldstein, and R.M. Newman. 2000. Relationship of wooded riparian zones and runoff potential to fish community composition in agricultural streams. *Canadian Journal of Fisheries and Aquatic Sciences* 57(2): 307-316.

Steinitz, C., H. Arias, S. Bassett, M. Flaxman, T. Goode, T. Maddock, D. Mouat, R. Peiser, and A. Shearer. 2003. Alternative futures for changing landscapes: The Upper San Pedro River Basin Arizona and Sonora. Island Press, Covelo, California.

Steinitz, C., R. Anderson, H. Arias, S. Bassett, M. Flaxman, T. Goode, T. Maddock, D. Mouat, R. Peiser, and A. Shearer. 2005. Alternative futures for landscapes in the Upper San Pedro River Basin of Arizona and Sonora. General Technical Report PSW-GTR-191. Forest Service, U.S. Department of Agriculture, Albany, California.

Story, A., R.D. Moore, and J.S. Macdonald. 2003. Stream temperatures in two shaded reaches below cutblocks and logging roads: downstream cooling linked to subsurface hydrology. *Canadian Journal of Forest Research* 33: 1,383–1,396.

Swales, S., and C. D. Levings. 1989. Role of off-channel ponds in the life cycle of coho salmon (Oncorhynchus kisutch) and other juvenile salmonids in the Coldwater River, British Columbia. 1989. *Canadian Journal of Fisheries and Aquatic Sciences* 46: 232–242.

Talmage, P.J., J.A. Perry, and R.M. Goldstein. 2002. Relation of instream habitat and physical conditions to fish communities of agricultural Streams in the northern midwest. *North American Journal of Fisheries Management* 22: 825–833.

Taylor, R.L., B.D. Maxwell, and R.J. Boik. 2006. Indirect effects of herbicides on bird food resources and beneficial arthropods. *Agriculture, Ecosystems & Environment* 116: 157-164.

Thomson, S.K., C.R. Berry, Jr., C.A. Niehus, and S.S. Wall. 2005. Constructed impoundments in the floodplain: A source or sink for native prairie fishes, in particular the endangered Topeka Shiner (Notropis topeka)? In: Glenn E. Moglen, editor, Watershed management 2005. Managing watersheds for human and natural impacts: engineering, ecological, and economic challenges. Proceedings of the 2005 Watershed Management Conference held in Williamsburg, VA, July 19-22, 2005. American Society of Civil Engineers, Reston, Virginia.

U.S. Environmental Protection Agency (EPA). 2004. The North Fork Potomac watershed story. EPA/903/F-04/002. Region III, Philadelphia, Pennsylvania.

U.S. Geological Survey (USGS). 2006. Pesticides in the nation's streams and ground water, 1992-2001. Circular 1291. Reston, Virginia. 172 pp.

Udawatta, R.P., J.J. Krstansky, G.S. Henderson, and H.E. Garrett. 2002. Agroforestry practices, runoff, and nutrient losses: a paired watershed comparison. *Journal of Environmental Quality* 31: 1,214-1,225.

Vannote, R.L., G.W. Minshall, K.W. Cummins, J.R. Sedell, and C.E. Cushing. 1980. The river continuum concept. *Canadian Journal of Fisheries and Aquatic Science* 37: 130-137.

Vesk, P.A., and R. MacNally. 2006. The clock is ticking : Revegetation and habitat for birds and arboreal mammals in rural landscapes of southern Australia. *Agriculture, Ecosystems and Environment* 112(4): 356-366.

Vidon, P.G.F. and A.R. Hill. 2004. Landscape controls on nitrate removal in stream riparian zones. *Water Resources Research* 40: 210-228.

Vondracek, B, K.L Blann, C.B. Cox, J.F. Nerbonne, K.G. Mumford, B.A. Nerbonne, L.A. Sovell, and J.K.H. Zimmerman. 2005. Land use, spatial scale, and stream systems: lessons from an agricultural region. *Environmental Management* 36: 775-791.

Wang, L., J. Lyons, P. Rasmussen, P. Seelbach, T. Simon, M. Wiley, P. Kanehl, E. Baker, S. Niemela, and P.M. Stewart. 2003. Watershed, reach, and riparian influences on stream fish assemblages in the Northern Lakes and Forest Ecoregion, U.S.A. *Canadian Journal of Fisheries and Aquatic Science* 60: 491–505.

Wehrly, K., M.J. Wiley, and P.W. Seelbach. 1998. Landscape-based models that predict July thermal characteristics of lower Michigan rivers. Fisheries Research Report No. 2037. Michigan Department of Natural Resources, Ann Arbor.

Wehrly, K.E., M.J. Wiley, and P.W. Seelbach. 2003. Classifying regional variation in thermal regime based on stream fish community patterns. *Transactions of the American Fisheries Society* 132: 18-38.

Weigel, B.M. 2003. Development of stream macroinvertebrate models that predict catchment and local stressors in Wisconsin. *Journal of the North American Benthological Society* 22: 123-142.

Weigel, B.M. J. Lyons, L.K. Paine, S.I. Dodson, and D.J. Undersander. 2000. Using stream macroinvertebrates to compare riparian land use practices on cattle farms in southwestern Wisconsin. *Journal of Freshwater Ecology* 15(1): 93-106.

Weller, D.E., T.E. Jordan, and D.L. Correll. 1998. Heuristic models for material discharge from landscapes with riparian buffers. *Ecological Applications* 8: 1,156-1,169.

Welsh, E.B., D.E. Spyridakis, J.I. Shuster and R.R. Horner. 1986. Declining lake sediment phosphorus release and oxygen deficit following wastewater diversion. *Journal of the Water Pollution Control Federation* 58: 92-96.

Welsh, H.H. Jr, G.R. Hodgson, and N.E. Karraker. 2005. Influences of the vegetation mosaic on riparian and stream environments in a mixed forest-grassland landscape in "Mediterranean" northwestern California. *Ecography* 28: 537-551.

Wildlife Management Institute. 2006. Lower Little Blue River watershed report: Data availability and monitoring of conservation policies and practices. Washington D.C. 27 pp.

Younus, M., M. Hondzo, and B.A. Engel. 2000. Stream temperature dynamics in upland agricultural watersheds. *Journal of Environmental Engineering* 126: 518-526.

Roundtable:
Realistic expectations of timing between conservation and restoration actions and ecological responses

Roundtable participants engaged in a wide-ranging discussion on many topics, most of them at least somewhat related to "realistic expectations." Among those topics were the following:

- Responses to environmental degradation are often technological fixes, but are the right end points being measured? The baseline of the "healthy" condition is often not known. Reversion to pristine conditions is impossible. "Recovery" is a healthy, diverse ecosystem, not native condition. The nitrogen cycle is distorted; the phosphorus cycle is broken; and hydrology has been altered. Balance and quality control are needed, but the economics does not work out. The Chesapeake Bay project was offered as an example.

- Ecological trajectories must be assessed to determine where they will lead in the future. The historical context is an important starting point from which to look forward and understand the trajectory of change. Factors must be assessed that cause changes in trajectories; what-if scenarios must be examined; and future scenarios from models must be developed. Population growth and pressure must be considered in these scenarios, along with climate change.

- How can people relate to realistic expectations? Realistically project or even come up with expectations? What limits what we can realistically expect or how we can change expectations? What directs the evolution of value systems through generations? Political will is needed to bring "realistic expectations" to reality, perhaps more than scientific or stakeholder interests.

- In considering expectations, the focus must be on progress—the right trajectory—rather than just end results.

- How should understanding of effects and expectations be scaled up from individual fields to entire watersheds?

- Agencies and other institutions continue to be data rich and information poor. Scientists could help by sorting out the key questions that might help turn existing data into useful information.

- Public involvement and sorting out what the public wants for the future is important. Community visioning processes and other exercises that help identify what is realistic and believable can help. The costs and behavioral changes involved need to be included in these processes.

- Policymakers must realize that conservation and restoration are long-term processes. Meaningful responses to conservation cannot be expected in the time frame of individual farm bills (five to seven years). Conservation effectiveness will require much longer time frames.

- Regional priorities must be defined that are meaningful to local farmers and populations. Environmental goals that are unrealistic and do not support reasonable integration of conservation and viable continuation of agricultural land use will not be accepted by farm operators.

- Monitoring of conservation effectiveness must be part of all conservation programs. A relatively small amount of high quality data can be used to extrapolate results to much larger areas and programs. But program managers must have the data, budgets, and long-term commitment to collect such information.

- Long-term monitoring of the effectiveness of agricultural conservation will require not only provision of financial support but development of an infrastructure that will support long-term collection of useable data and results. This will require setting measurable, reasonable goals and identification of an agency responsible for training, data quality control, interpretation of results, and getting those results to the public and people who make long-term agricultural policy decisions (U.S. Department of Agriculture officials and political representatives in Washington, D.C).

Roundtable participants then reached consensus on a series of leading questions that at least implied what the most important next steps might be in strengthening the science important to agricultural conservation:

1. How do we identify reasonable expectations? How do we communicate them to the public and policymakers? How do we receive communications back from the public and policymakers? How do we make adaptive management work in the "real world," that is, how do we involve the public in adaptive management (and who are "we")?

2. How can we develop a process to identify and influence trajectories of change and do so at an ecosystem/landscape level rather than a localized, single-issue level? What are the costs and benefits of alternative trajectories? There are many measures to assess in evaluating alternative future scenarios. A process for doing this has been used in some areas, but is not widely available or widely known.

3. What questions do we need to ask and answer to turn data into information that can be used to refine realistic expectations? Where do we need more data, and where do we just need to analyze what we have?

4. What is an appropriate timeframe in which to develop reasonable expectations? What are people's/politicians' typical timeframes? How do "realistic" expectations change when the time frame is 2 to 4 years, 10 years, a generation, 100 years, or more?

5. Realistic expectations are subject to change over time. What factors, both catastrophic and evolutionary, cause perceptions of what is realistic to change? What can we do to avoid being only passive participants in this process?

The promise of adaptive management

Mary C. Watzin

The challenge of managing agricultural land with both productivity and environmental integrity in mind is not new. Since the earliest days in the history of the United States, agriculture has been a vital part of the economy, culture, and character of the landscape. Even though farms in the earliest days of the history of the United States were small and used primarily to support families and local communities, the environmental impacts of land-clearing and other practices were apparent and the need for conservation debated. While soil and water conservation practices accomplished much in the twentieth century, the cumulative effects of intensive and mechanized agriculture, huge increases in fertilizer and chemical usage, animal confinement, and other more recent farming practices have led to widespread declines in the ecological integrity of agricultural landscapes. Recently, the U.S. Environmental Protection Agency (2002) estimated that no less than 40 percent of the nation's rivers, lakes, and estuaries were impaired. Many of those systems were impaired by the cumulative effects of agriculturally derived nonpoint-source pollution.

The 2002 farm bill dramatically increased federal investment in conservation practices across agricultural landscapes – and the expectation that improvements in water quality and the environment would result from those investments. This expectation included not just the hope that improvements might be seen on agricultural fields and at the edges of those fields, but also that improvements would begin to emerge at the landscape or watershed scale and that the list of impaired waters would begin to shrink. Across the country, many efforts are underway to try to quantify the benefits of applying conservation practices on agricultural land at the watershed scale, and changes have been difficult to discern. As a result, a number of questions have arisen about what expectations managers and policymakers should have about the time between conservation implementation and resulting environmental improvements.

Gregory and colleagues, in their earlier chapter in this book, argued that because conservation practices, by definition, are attempting to "protect" existing ecosystem structure and function, outcomes are expected to be relatively rapid. In contrast, they suggest that a much longer timeline is expected for restoration practices, which are designed to "bring back" a portion of the ecosystem structure and function that has been lost. For restoration, the desired ecosystem outcomes might not be attained for decades or longer.

For agricultural conservation programs, however, it is also important to recognize the difference between practice design and application. With the obvious exception of land retirement programs, most conservation practices we employ are designed to minimize the adverse environmental impacts of ongoing agriculture and protect existing environmental goods and services. We hope they will also result in improved environmental conditions over time. By design, those practices are not restoration, that is, they were not developed primarily to fix significant environmental problems already in existence across the landscape. In application, however, we are using conservation practices as restoration – we are applying them in watersheds where water quality impairments are already significant, and we are expecting to see environmental improvements occur in a relatively short period of time. It is probably not realistic to expect that practices designed to protect and maintain, or even to minimize harm, will result in rapid ecological improvements across the landscape.

In order to develop realistic expectations about improvements in impaired agricultural watersheds, we must commit to understanding how individual improvements accumulate across space and time. While a number of studies provide insights about the effects of conservation practices at the edge of the farm, we have much less information about how improvements accumulate across landscapes and watersheds. We are also just beginning to understand the lag times that should be expected in natural systems because of the build-up of nutrients, changes in hydrology and geomorphology, and myriad other alterations that have accumulated in altered ecosystems. In the face of these unknowns, the rate at which we achieve our restoration and conservation goals will be directly proportional to how fast we learn and apply what we have learned.

Adaptive management

Adaptive management is a systematic process for continually improving management policies and practices by learning from the outcomes of practices and programs that are underway. Adaptive management was developed in the 1970s (Holling, 1978; Walters, 1986; Lee, 1993) in part as a method for dealing with uncertainty, acknowledging that because scientific understanding of ecosystems and the potential outcomes of management actions are incomplete, we should learn as we go. It is an information-based, iterative approach based on structured learning.

Adaptive management explicitly incorporates experimentation into the design and implementation of management policies (Holling, 1978; Noss et al., 1997; Lee, 1999; Schreibner et al., 2004; Stankey et al., 2005). By adopting an adaptive management approach, policymakers acknowledge that their actions may be modified over time as better information becomes available. Lack of knowledge should not be an excuse for inaction, but the limits of knowledge and the assumptions underlying actions must be clearly acknowledged.

In adaptive management, an action plan is designed on the basis of best current professional judgment. Actions are then implemented as tests of hypotheses or predictions about how managers and scientists think the ecosystem is working. As implementation occurs, monitoring data are collected. As data accumulate, they are evaluated to

judge the effectiveness of the actions in achieving the desired outcomes. If outcomes are not being achieved as expected, then adjustments are made in the action plan, and the cycle begins again (Figure 1).

If data are systematically collected before and after implementation, changes in monitored indicator levels can provide information about the condition of the ecosystem, the effectiveness of the management actions taken, and the validity of the hypotheses upon which the management approach was developed. Using an adaptive management approach, managers have the opportunity to increase their understanding of the ecosystem, regardless of the outcomes of management. This increases the potential success of future management strategies.

The ability of adaptive management to support good decision-making, therefore, is directly dependent upon the quality of the monitoring program. While monitoring for individual practices can be conducted at the plot or field scale, monitoring for whole-farm conservation systems, such as Environmental Quality Incentives Program (EQIP) farm plans, or for documenting changes in impaired agricultural watersheds, must be conducted at the landscape scale to determine the combined outcomes of all practices. The indicators selected for the monitoring program must include measures that can detect changes in water quality and other valued environmental components following installation of conservation practices; measures are needed also that can help relate changes in these ecosystem components to the presence of conservation practices.

The elements of such a monitoring program must include:

- Baseline monitoring, to document the current state of the resource.
- Implementation monitoring, to document that practices were implemented as intended.
- Effectiveness monitoring, to determine whether desired outcomes are achieved.
- Validation monitoring, to determine if the original assumptions and hypotheses that drove the selection of practices and their placement across the watershed were correct.

At the watershed scale, baseline monitoring should document the impairments, the pollutants or habitat alterations that are causing the impairments, and the critical source areas for the pollut-

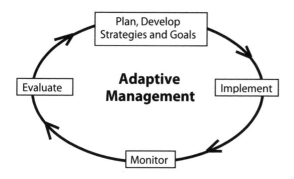

Figure 1. The adaptive management cycle.

ants. Because of the strong relationship between water flow, soil type, and slope and pollutant generation, these covariates should also be measured. For water flow, this will require monitoring across a sufficient number of storm events.

In order to relate changes in environmental conditions to the implementation of conservation practices, monitoring design (or experimental design) demands considerable attention. For water quality concerns, the longer the duration of monitoring, the more likely it is that changes in pollutant loading and stream condition might be detected because of large year-to-year variability in rainfall and runoff patterns. For watersheds where background variations in these parameters are known, sensitivity analyses can help determine the sample frequency and duration necessary to provide reasonable resolution (Richards and Grabow, 2003).

Farm practice monitoring must also be included in the adaptive management monitoring program because changes in animal numbers, cropping patterns, and the like might influence outcomes. Conservations practices must be implemented on a sufficient number of farms throughout a watershed to ensure documentation of the effects of those practices on water quality and the environment if changes occur.

While various programs have been monitoring in watersheds across the United States for decades for different purposes (e.g., the U.S.Geological Survey's National Water Quality Assessment Program and the U.S. Environmental Protection Agency's Section 319 monitoring program), the real value and promise of adaptive management is in its proactive approach. The focus in adaptive management is on providing information to

guide decision-making in the future, not simply to document successes and failures in the past. For decades, we have conducted a large-scale experiment across the landscape without collecting the data we need to interpret it. Clearly, adaptive management could provide a context for addressing these unknowns if a set of indicators is adopted and the monitoring data to quantify them are collected over time.

Using adaptive management in conservation programs

Adaptive management has primarily been used by management agencies for large-scale restoration projects, such as those in the Everglades (U.S. Army Corps of Engineers, 1999; National Research Council, 2003), the Glen Canyon section of the Colorado River (National Research Council, 1999), and the Columbia River Basin (Lee, 1993). Only more recently have professionals considered applications of adaptive management in an agency setting as a way to improve the effectiveness of ongoing programs (Johnson, 1999; Moir and Block, 2001; Freedman et al., 2004; Harris and Heathwaite, 2005).

Within the context of agricultural conservation programs, adaptive management might best be applied by focusing on classes of problems that require similar types of decisions in different situations and locations. For each class of problems, general guidelines might be articulated that serve as hypotheses about which conservation practices might be applied and the improved outcomes that might result, with modifications to account for site-specific characteristics and limitations. A monitoring program could then be established to document progress towards those expected outcomes and evaluate the guidelines.

This monitoring program should take advantage of data that are already being collected by farmers and other agricultural stakeholders. For example, an essential part of comprehensive nutrient management is collecting soil phosphorus concentration or index values. In addition to providing information about the nutritional status of the soil, these data can also indicate the potential for losses that can negatively affect water quality, especially when combined with a soil erosion indicator. For waters listed as 'impaired" on the U.S. Environmental Protection Agency (EPA)

303(d) list and subject to total maximum daily load (TMDL) requirements, states often have considerable water quality and other information assembled to support the preparation of a load allocation. Ninety percent of TMDL sites nationwide involve nonpoint sources; those restoration plans could benefit from a watershed approach (U.S. Environmental Protection Agency, 2005) and from data that might be gathered through an adaptive management approach.

Within the Natural Resources Conservation Service (NRCS), the conservation system guides and customer service toolkit provide an opportunity to assist in gathering information that might be used to evaluate the effectiveness of conservation practices across the country. Resource management systems are already recommended by common resource area and land use. For each group of practices, there are baseline descriptions and hypothesized estimates of system effects for various resource concerns. As quantitative estimates for parameters are added to the guides, expected outcomes can be compared to results documented on the ground. Because the NRCS customer service toolkit is linked to the conservation system guide and georeferenced, outcomes could be accumulated by drainage area as conservation practices are implemented within watersheds.

To work most effectively, however, indicators that relate to the conservation practices must be chosen up front. For each indicator, a measurement schedule should be established, the level of each indicator that reflects an acceptable outcome should be articulated, and the timeframe in which this change might be expected should be estimated. Establishing these levels and timeframes will require open communication between scientists, managers, stakeholders, and the public. A consultation and decision plan should be established that defines how information will be shared, when initial judgments about adequate progress might be made, and how actions might be modified if progress is unacceptable.

The lack of such a decision plan is one reason why adaptive management has not reached its full potential. Plans are prepared, indicators selected, and monitoring data collected, but a timetable for reevaluation is not established, performance targets or acceptable levels are not established for the indicators (Smyth et al., in press), and management responses that might be

taken if progress is not acceptable are not considered in any explicit way. Construction of a monitoring table (Table 1) that outlines those things at the beginning of an adaptive management application could help in guiding development of the decision plan. Schedules for decision-making first might focus on assessing implementation using program measures, then move to system responses. Acceptable levels might be established as trajectories of change if the timeframes and thresholds of response are too uncertain, but the more definitive the monitoring plan is, the more likely that it will lead to completion of the adaptive management cycle and more proactive management over the long term.

Certainly, understanding impacts at the edge of the field or farmstead is an easier proposition than understanding how effects accumulate across whole landscapes or watersheds. There are numerous examples across the country of implementation of conservation practices on individual sites, with little or no change in water quality at the watershed level. This is likely the result of many factors, some controllable, some uncontrollable. Included are weather, lack of targeting of critical source areas, insufficient percentage of farms participating in the action plan, ineffective practices or level of treatment, and cumulative changes in the biophysical processes in the watershed. Adaptive management offers the opportunity to explore these contributing factors and fine-tune the management approach. This includes bringing partners with diverse tools for management into the overall effort.

Recently, Gregory et al. (2006) proposed a decision framework for assessing the merits of a proposed application of adaptive management. They suggested that adaptive management has its greatest value when it reduces uncertainty by introducing new information that has the potential to change management decisions and efficiencies and thereby produce gains in objectives that collective stakeholders value. In developing an adaptive management approach, they also recommend considering the tradeoffs involved in incorporating existing information, best professional judgment, and new experimentation into the overall strategy. Similarly, both Richards (2004) and Harris and Heathwaite (2005) advocated use of an approach that combines modeling with empirical data collection to cope with uncertainty

Table 1. Suggested structure for developing an adaptive management monitoring plan. Indicators should be selected and developed within each of the four categories.

Indicator	Measurement schedule	Acceptable level	Management response if unacceptable
Program measures			
Physical/chemical measures			
Biological measures			
Socioeconomic measures			

and develop more parsimonious predictions that might inform adaptive management. The recent blue ribbon panel on CEAP—Conservation Effects Assessment Project—convened by the Soil and Water Conservation Society (SWCS), advocated an adaptive approach that would combine modeling, on-the-ground monitoring, process-oriented research in experimental watersheds, and a proactive planning and management framework that compared progress to established goals (Soil and Water Conservation Society, 2006).

Watzin et al. (2003) argued that there is considerable merit in merging a focused, on-the-ground monitoring program with a whole-farm mass balance model that is linked to a spatially explicit landscape model. The landscape model could capture watershed hydrology and the links between hydrologic processes and water quality. By using the whole-farm mass balance model to capture realistic farm operations and long-term trends in nutrient dynamics, and then coupling this output to the larger landscape model, watershed-level improvements accumulating from implementation on a number of farms in the watershed could be predicted. The landscape model might also be used to examine the ecological and economic impacts of both existing and proposed conservation practices under farm bill programs.

Use of models to explore multiple future scenarios can help cope with uncertainties and incorporate a variety of alternative perspectives and possibilities into conservation planning and implementation (Peterson et al., 2003). Outcomes can include both economic and ecological considerations. For example, Wu et al. (2004) presented

a model that combines farm-based cropping and conservation practice information with a regional agricultural land use model. The model simulates the payments for conservation tillage under each scenario and combines these with simple estimates of soil and nitrogen losses. Using the outputs from each scenario, decision-makers can explore the most cost-effective policies for reducing soil and nutrient loading to downstream receiving waters.

An adaptive management approach might also be used to explore whether our fundamental policy approaches effectively meet our environmental goals. Each of our conservation programs and practices has different goals and different approaches, which sometimes conflict with one another. For example, conservation programs that provide compensation for retiring land might compete with programs available only for actively farmed land. The structure of our programs may also slow environmental progress. For example, federal programs support development of manure management systems that turn dairy waste into a liquid product that is not easily transported out of local areas. A large land base is required for disposal as a result. In Vermont there is simply not enough cropland to make appropriate use of this product, and high soil phosphorus concentrations are common across the agricultural landscape (Magdoff et al., 1997; Jokela et al., 2004). While alternative practices, such as liquid and solid separation and composting, might reduce the soil phosphorus problem, it is more difficult for producers to implement these practices under current programs.

As a nation, we have also taken a very different policy approach to managing the environmental impacts of industry than we have taken to managing the environmental impacts of agriculture. Instead of using regulations and fines, with the exception of confined animal feeding operation (CAFO) requirements, we have adopted a largely voluntary program to manage agricultural impacts, one that relies on financial incentives and the actions of willing participants. This has made it more challenging to target interventions and optimize investments across the landscape. If we also consider that state governments and an ever-increasing number of nongovernmental organizations are also investing in conservation and restoration, the most efficient and effective action plan becomes even more challenging – but also has enormous potential. Adaptive management provides a framework for comparing untargeted and targeted futures and optimizing the type and the placement of practices across the landscape to achieve desired outcomes.

Using existing data to guide an adaptive management approach

A large literature base exists to help guide development of an adaptive management action plan for agricultural landscapes. Although most of these studies examine effects at the plot scale or at the edge of the farm, in the immediate vicinity of implementation, some studies can appropriately be used to extrapolate results to the larger landscape or watershed scale.

The paired-watershed approach, in which two similar drainage basins are monitored before and after implementation of conservation practices in one of the drainage basins, is one of the more powerful approaches for documenting effectiveness over a larger area. Although the number of paired-watershed studies conducted in agricultural watersheds is not high, data from these studies provide useful information about the timeframes of response and some factors that might drive these timeframes.

For example, a paired-watershed study in the drainage of Cannonsville Reservoir, which supplies drinking water for New York City, showed that intensive management, including manure management, rotational grazing, and improved infrastructure, resulted in significant reductions in both dissolved and particulate phosphorus within four years of implementation (Bishop et al., 2005). In the Lake Champlain Basin, implementation of livestock exclusion in an experimental watershed resulted in significant reductions in both phosphorus load and fecal coliforms in just three months (Meals, 2001).

In contrast, 10 years of monitoring the St. Albans Bay watershed in Vermont as part of the Rural Clean Water Program failed to document any significant changes in water quality in the bay, despite a combination of both wastewater treatment plant upgrades and implementation of dairy waste management practices (Meals, 1992). In this case, the lack of effect is likely the result of the high residual load of phosphorus in the sediments, which continue to supply dissolved phosphorus to the overlying water (Druschel, 2005).

Similarly, a 15-year study of riparian buffers established along streams in the Stroud Preserve, Pennsylvania, showed no reduction in nitrate movement across the buffer for the initial 10-year period, reflecting, in large part, the time needed for the vegetation to become established (Szpir et al., 2005). Another paired-watershed study of riparian buffers also documented the long period necessary for plants to become established (Udawatta et al., 2002), but demonstrated that buffers can reduce the pollution entering adjacent streams if they intercept major overland flowpaths. The effectiveness of nutrient removal in an established buffer often decreases with time, however, when nutrients accumulate in the flow paths (Sheppard et al., 2006). Other studies also suggest that filter strips and buffers can become saturated with phosphorus and no longer effectively remove nutrients over time, especially if they are used as the primary method of treatment (Dillaha et al., 1989; Reed and Carpenter, 2002; Watzin et al., 2003). A recent review of conflicting data about riparian buffers notes that spatial variation in local conditions often explains why some buffers show high pollutant removal efficiencies while others do not (Polyakov et al., 2005); these authors suggest that precise design of buffers using this information might greatly improve their ecological effectiveness and their economic viability. This is exactly the kind of analysis and revised implementation approach that adaptive management can foster.

Significant unknowns

Despite the examples of what we know about the performance of some conservation practices, numerous important unknowns remain. Those unknowns make prediction of effects of practice implementation and timeframes of response uncertain. Several important examples include:

- Surface and shallow subsurface flowpaths, in both drained and undrained fields, which can dramatically affect hydrologic transport of pollutants.
- Nutrient concentration and dynamics in streambank and riparian soils and the potential for those soils to contribute nutrients to streams.
- Controls on spatial variations in soil phosphorus concentrations (and soil test measurements) and their response to nutrient management, including lag times in response.
- The relative contribution of direct farmstead discharges (milkhouse waste, barn and barnyard waste, and silage leachate) to the watershed pollutant load.
- Thresholds of treatment necessary (or level of practice implementation) to affect a stream reach or downstream receiving water.
- The relative contribution of various conservation practices to improved water quality at the watershed scale, for example, the contribution of livestock exclusion from streams versus filtering up-slope runoff.

One significant, unanswered question in river and watershed science and management is the extent to which the eroding banks and beds of actively adjusting stream and river channels contribute to sediment and nutrient loads in impaired waters. Several studies show that streams in agricultural landscapes have increased bank erosion and deeper bottom sediments (Dovciak and Perry, 2002; Townsend et al., 2004; Allan, 2004a) and that land use impacts operate across a variety of scales (Watzin and McIntosh, 1999; Allan, 2004b; Stallins, 2006). On a watershed scale, there is increasing evidence that the sediment and phosphorus generated by fluvial processes in adjusting streams may overshadow or overwhelm the landscape-derived nonpoint-source inputs of these pollutants (Richards et al., 1996; Allan, 2004a,b).

Once in the stream, phosphorus and its associated sediments are distributed by fluvial processes. Those processes are a function of the links between stream reaches and river-system dynamics (Gburek et al., 2000; McDowell et al., 2003). Stream reaches in poor geomorphic condition can either store or transport soil-associated phosphorus in the reach. In particular, reaches that are degrading or incising will transport sediment and phosphorus downstream.

Reach-level geomorphic data provide a wealth of information about channel form, sediment accumulation, and sediment erosion at specific locations. But these data are not currently linked to provide either a longitudinal (headwaters to river mouth) or watershed-wide perspective on sediment and nutrient dynamics. Adaptive management could provide a context for linking reach-level geomorphic data with information about sediment phosphorus or other pollutant concentrations in agricultural watersheds; it could also provide a unique opportunity to understand the contribution of both fluvial (instream-eroded) sediment and land-based pollutants to water quality impairments. Better understanding of the transport of pollutants from landscapes to rivers, and ultimately to downstream receiving waters, might allow managers to target conservation practices and other restoration measures to those reaches contributing most to the pollutant load. Use of geomorphic assessment has become widespread in stream management (Rosgen, 1996; Montgomery and Buffington, 1997; Federal Interagency Stream Restoration Working Group, 1998). If a link between geomorphic condition and water quality exists, managers may be able to use geomorphic data not only to manage river corridors and protect roads and bridges, but also to make informed decisions about actions needed to restore and protect water quality.

Gradient analysis and geostatistical techniques offer the potential to explain patterns of variation across environmental variables and spatial scales (Braak and Prentice, 1988; Isaaks and Srivastava, 1989; Jager et al., 1990; Goovaerts, 1994). Decision trees can be used to examine iteratively management options and system-specific circumstances. A decision-tree approach might help link processes, responses, and boundary conditions in logical sequences that can support decision-making by managers (Tan et al., 2005). By simultaneously examining land use, soil pollutant concentrations, hydrologic connections, geomorphic condition, and discharge at the farm and whole watershed scales, a greater understanding of the relative importance of both local and larger scale factors in controlling stream water quality and ecological integrity might develop. Clearly, if these relationships are not understood, it will not be possible, realistically, to predict risks to water quality, or likely improvements after conservation practices are implemented (Pionke et al., 2000; Heathwaite et al., 2003; Page et al., 2005).

In the Lake Champlain Basin, increasing land use changes within both agricultural and urban watersheds are associated with declining geomorphic condition and ecological integrity in stream channels. Stream reaches in poor geomorphic condition frequently have high rates of bank failure or channel incision. In the last 15 years, the sedimentation rate in Missisquoi Bay, which drains a large agricultural watershed in the Lake Champlain Basin, has increased by an order of magnitude, going from less than 0.1 centimeter (0.04 inch) to more than 1 centimeter (0.4 inch) per year (Princhonnet, 2003). It is possible that in heavily altered watersheds a threshold of stream adjustment can be exceeded, such that increasingly longer river stretches are exporting sediment and phosphorus to receiving waters, thus increasing the time it will take for these watersheds to respond to the implementation of conservation practices.

Another large area of uncertainty surrounds these same kinds of questions. Implementation of conservation practices across the landscape will not result in uniform and linear responses in water quality. The placement of interventions on the landscape, soils, topography, hydrology, and other factors all influence how a watershed responds to conservation treatments. The growing literature on ecological thresholds and nonlinear changes can help scientists and managers understand the relative contribution of landscape context, hydrologic source areas, and geomorphic instability (stream channel adjustment) to resource degradation and the nature and timeframes of the response to conservation practice implementation.

The literature on ecological thresholds suggests that there are breakpoints in ecosystem processes that, once exceeded, can trigger major changes in other ecosystem components (Scheffer et al., 1993, 2001; Paine et al., 1998; Rapport and Whitford, 1999; Muradian, 2001; Church, 2002; Carpenter,

2005). Ecological thresholds might be envisioned as the transition point between an acceptable level of anthropogenic stressors, one to which the ecosystem can adapt, and an unacceptable level of these stressors, one that drives the ecosystem to change to an undesired state. Unfortunately, ecological thresholds frequently only become apparent after disastrous shifts in community composition have occurred; therefore, managers have not generally succeeded in preventing catastrophic shifts in state.

Ecosystems in good condition feature resilience; they maintain the ability to recover from both natural disturbances and some human perturbations (Holling, 1973). This resilience depends upon a variety of self-reinforcing mechanisms or feedbacks that prevent the ecosystem from shifting to an alternative stable state. It can also be considered the adaptive capacity of the system. Increasing evidence suggests that ecosystem resilience may be among the most important characteristics in place to sustain the human uses and ecosystem services that a growing human population – and agricultural enterprise – requires (Gunderson and Holling, 2002; Elmqvist et al., 2003).

Recently, Gunderson and Holling (2002) proposed an approach to natural resource management based on evaluation of the types and timing of management interventions in light of their ability to increase resilience. Walker et al. (2002) built on this approach and proposed a four-step process for resilience management. By increasing understanding of resilience, the goal is to identify actions that will strengthen the feedback relationships that maintain the desired state of the ecosystem. One of the challenges in developing sustainable management and recovery strategies, however, is that both the feedback mechanisms and the relationships between feedback mechanisms and spatial scale are poorly understood (Rietkerk et al., 2004). Using a combination of modeling and data review, Carpenter et al. (1999) showed that in some lakes cultural eutrophication might be reversed by reducing phosphorus loading alone, but in other lakes a threshold had been crossed and the changes were likely irreversible if phosphorus reduction was the only management approach taken. If inputs of phosphorus and other water pollutants are stochastic, as they are from agricultural nonpoint sources, and there are uncertainties about the mechanisms of response, then a high likelihood exists that nonlinear changes will occur and response times to management interventions will be longer and more difficult to predict.

Summary and conclusions

Perhaps the greatest strength of adaptive management is its holistic nature. The many cumulative effects of implementing conservation practices across the landscape cannot be separated. They must be considered collectively. Adaptive management will force an ordered and systematic consideration of the range of practice effects. The approach provides a framework and an opportunity to look beyond on-farm, direct effects and consider the broader ecological consequences of implementing a suite of conservation practices. It can allow a more proactive, problem-solving approach to achieving desired ecological outcomes.

Adaptive management is not a "make-it-up-as-you-go" approach (Stankey et al., 2005). Neither is it just a conceptual ideal. It demands a deliberative commitment to structured learning. The only path to better predictions about realistic timeframes of response is through improved conservation practices and improved ecological understanding of how landscape processes modify and control the ecological outcomes of using those conservation practices.

By exploring scenarios that examine different combinations and placements of conservation practices within a watershed and monitoring the pattern of response across the landscape, we can determine which combinations and placements have the greatest potential for improving ecological outcomes. We know that lag times exist. Only a commitment to adaptive management, with regular reevaluation and modification of our action plans, will reduce these lag times and achieve improvements in agricultural landscapes on the shortest timeline possible.

References

Allan, J.D. 2004a. Influence of land use and landscape setting on the ecological status of rivers. *Limnetica* 23: 187-198.

Allan, J.D. 2004b. Landscape and riverscapes: the influence of land use on river ecosystems. *Annual Reviews of Ecology, Evolution and Systematics* 35: 257-284.

Bishop, P.L, W.D. Hively, J.R. Stedinger, M.R. Rafferty, J.L. Loipersberger, and J.A. Bloomfield. 2005. Multivariate analysis of paired watershed data to evaluate agricultural best management practice effects on stream water phosphorus. *Journal of Environmental Quality* 34: 1,087–1,101.

Braak, C.J.F. and I.C. Prentice. 1988. A theory of gradient analysis. *Advances in Ecological Research* 18: 271-313.

Carpenter, S.R. 2005. Eutrophication of aquatic ecosystems: bistability and soil phosphorus. *Proceedings of the National Academy of Sciences* 102: 10,002-10,005.

Carpenter, S.R., D. Ludwig, and W.A. Brock. 1999. Management of eutrophication for lakes subject to potentially irreversible change. *Ecological Applications* 9: 751-771.

Church, M. 2002. Geomorphic thresholds in riverine landscapes. *Freshwater Biology* 47: 541- 557.

Dillaha, T.A., R.B. Reneau, S. Mostaghimi, and D. Lee. 1989. Vegetative filter strips for agricultural nonpoint source pollution control. *Transactions, American Society of Agricultural Engineers* 32(2): 513-519.

Dovciak, A.L. and J.A. Perry. 2002. In search of effective scales for stream management: does agroecoregion, watershed or their intersection best explain the variance in stream macroinvertebrate communities? *Environmental Management* 30: 365-377.

Druschel, G.K., A. Hartmann, R. Lomonaco, and K. Oldrid. 2005. Determination of sediment phosphorus concentrations in St. Albans Bay, Lake Champlain: assessment of internal loading and seasonal variations of phosphorus sediment-water column cycling. Vermont Agency of Natural Resources, Montpelier. 71 pp.

Elmqvist, T., C. Folke, M. Nystrom, G. Peterson, J. Bengtsson, B. Walker, and J. Norberg. 2003. Response diversity, ecosystem change, and resilience. *Frontiers in Ecology and the Environment* 1(9): 488-494.

Federal Interagency Stream Restoration Working Group. 1998. Stream corridor restoration: principles, practices and processes. U.S. Department of Agriculture, Washington, D.C.

Freedman, P.L., A.D. Nemura, and D.W. Dilks. 2004. Viewing Total Maximum Loads as a process, not a singular value: adaptive watershed management. *Journal of Environmental Engineering* 130(6): 695-702.

Gburek, W.J., A.N. Sharpley, L. Heathwaite, and G.J. Folmar. 2000. Phosphorus management at the watershed scale: a modification of the Phosphorus Index. *Journal of Environmental Quality* 29: 130–144.

Goovaerts, P. 1994. Study of spatial relationships between two sets of variables using multivariate geostatistics. *Geoderma* 62(1-3): 93-107.

Gregory, R., L. Failing, and P. Higgins. 2006. Adaptive management and environmental decision-making: a case study application to water use planning. *Ecological Economics* 58(2): 434-447.

Gunderson, L. and C.S. Holling, editors. 2002. *Panarchy: understanding transformations in human and natural systems.* Island Press, Washington, D.C.

Harris, G. and A.L. Heathwaite. 2005. Inadmissible evidence: knowledge and prediction in land and riverscapes. *Journal of Hydrology* 304: 3-19.

Heathwaite, A.L., A.I. Fraser, P.J. Johnes, M. Hutchins, E. Lord, and D. Butterfield. 2003. The Phosphorus Indicators Tool: a simple model of diffuse P loss from agricultural land to water. *Soil Use Management* 19(1): 1–11.

Holling, C.S. 1973. Resilience and stability of ecological systems. *Annual Review of Ecology and Systematics* 4: 1-23.

Holling, C.S., editor. 1978. *Adaptive environmental assessment and management.* John Wiley, London, England. 377 pp.

Isaaks, E.H. and R.M. Srivastava. 1989. *Applied geostatistics.* Oxford University Press, New York, New York.

Jager H.I., M.J. Sale, and R.L. Schmoyer. 1990. Cokriging to assess regional stream quality in the southern Blue Ridge Province. *Water Resources Research* 26(7): 1401-1412.

Johnson, B.L. 1999. The role of adaptive management as an operational approach for resource management agencies. *Conservation Ecology* 3(2): 8.

Jokela, W.E., J.C. Clausen, D.W. Meals, and A.N. Sharpley. 2004. Effectiveness of agricultural best management practices in reducing phosphorus loading to Lake Champlain. In T. Manley, P. Manley, and T. Mihuc, editors, *Lake Champlain: partnerships and research in the new millennium.* Kluwer Academic/Plenum Publishers, New York, New York.

McDowell, R.W., A.N. Sharpley, and G. Folmar. 2003. Modification of phosphorus export from an eastern USA catchment by fluvial sediment and phosphorus inputs. *Agriculture Ecosystems and Environment* 99: 187-199.

Lee, K.N. 1993. *Compass and gyroscope: integrating science and politics for the environment.* Island Press, Washington, D.C. 243 pp.

Lee, K.N. 1999. Appraising adaptive management. *Conservation Ecology* 3(2): 3.

Magdoff, F., L. Lanyon, and B. Liebhardt. 1997. Nutrient cycling, transformations, and flows: implications for a more sustainable agriculture. *Advances in Agronomy* 60: 1-73.

Meals, D.W. 1992. Water quality trends in the St. Albans Bay, Vermont watershed following RCWP land treatment. In The National Rural Clean Water Program Symposium, September, 1992, Orlando, Florida. EPA/625/R-92/006. U.S. Environmental Protection Agency, Washington, D.C.

Meals, D.W. 2001. Lake Champlain Basin agricultural watersheds Section 319 national monitoring program project, final project report: May 1994-September, 2000. Vermont Department of Environmental Conservation, Waterbury. 227 pp.

Moir, W.H. and W.M. Block. 2001. Adaptive management on public lands in the United States: commitment or rhetoric? *Environmental Management* 28(2): 141–148.

Montgomery, D.R. and J.M. Buffington. 1997. Channel-reach morphology in mountain drainage basins. *Geological Society of America Bulletin* 109(5): 596-611.

Muradian, R. 2001. Ecological thresholds: a survey. *Ecological Economics* 38(1): 7-24.

National Research Council. 1999. Downstream: adaptive management of Glen Canyon Dam and the Colorado River ecosystem. Committee on Grand Canyon Monitoring and Research, National Research Council, Washington, D.C. 242 pp.

National Research Council. 2003. Adaptive monitoring and assessment for the comprehensive Everglades restoration plan. Committee on Restoration of the Greater Everglades Ecosystem, National Research Council, Washington D.C. 122 pp.

Noss, R.F., M.A. O'Connell, and D.D. Murphy. 1997. *The science of conservation planning: habitat conservation under the Endangered Species Act.* Island Press, Washington, D.C.

Page, T., P.M. Haygarth, K.J. Beven, A. Joynes, T. Butler, C. Keeler, J. Freer, P.N. Owens, and G.A. Wood. 2005. Spatial variability of soil phosphorus in relation to the topographic index and critical sources areas: sampling for assessing risk to water quality. *Journal of Environmental Quality* 34(6): 2,263-2,277.

Paine, R.T., M.J. Tegner, and E.A. Johnson. 1998. Compounded perturbations yield ecological surprises. *Ecosystems* 1: 535-545.

Peterson, G.D., G.S. Cumming, and S.R. Carpenter. 2003. Scenario planning: tool for conservation in an uncertain world. *Conservation Biology* 17(2): 358-366.

Princhonnet, G. 2003. Caractérisation des sédiments de la baie Missisquoi et de la rivière-aux-Brochets (C, N, P et métaux lourds). Université du Québec à Montréal, Département des sciences de la Terre et de l'atmosphère (SCT-SCA) GEOTERAP Project Report. 83 pp.

Pionke, H.B., W.J. Gburek, and A.N. Sharpley. 2000. Critical source area controls on water quality in an agricultural watershed located in the Chesapeake Basin. *Ecological Engineering* 14(4): 325–335.

Polyakov, V., A. Fares, and M.H. Ryder. 2005. Precision riparian buffers for the control of nonpoint source pollutant loading into surface waters: a review. *Environmental Reviews* 13(3): 129-144.

Rapport, D.J., and W.G. Whitford. 1999. How ecosystems respond to stress? Common properties of arid and aquatic systems. *Bioscience* 49(3): 193-203.

Reed, T., and S.R. Carpenter. 2002. Comparisons of P-yield, riparian buffer strips, and land cover in six agricultural watersheds. *Ecosystems* 5(6): 568-577.

Richards, R.P. 2004. Improving Total Maximum Daily Loads with lessons learned from long-term detailed modeling. *Journal of Environmental Engineering* 130 (6): 657-663.

Richards, R.P. and G.L. Grabow. 2003. Detecting reductions in sediment loads associated with Ohio's Conservation Reserve Enhancement Program. *Journal of the Water Resources Association* 39(5): 1,261-1,268.

Richards, C., L.B. Johnson, and G.E. Host. 1996. Landscape-scale influences on stream habitats and biota. *Canadian Journal of Fisheries and Aquatic Sciences* 53(suppl. 1): 295-311.

Rietkerk, M., S.C. Dekker, P.C. de Ruiter, and J. van de Koppel. 2004. Self-organized patchiness and catastrophic shifts in ecosystems. *Science* 305: 1,926-1,929.

Rosgen, D.L. 1996. Applied river morphology. Wildland Hydrology, Pagosa Springs, Colorado.

Scheffer, M., S.H. Hosper, M. L. Meijer, B. Moss, and E. Jeppesen. 1993. Alternative equilibria in shallow lakes. *Trends in Ecology & Evolution* 8(8): 275-279.

Scheffer, M., S. Carpenter, J. Foley, C. Folke, and B. Walker. 2001. Catastrophic shifts in ecosystems. *Nature* 413: 591-596.

Schreibner, E.S.G., A.R. Bearlin, S.J. Nicol, and C.R. Todd. 2004. Adaptive management: a synthesis of current understanding and effective application. *Ecological Management and Restoration* 5: 177-182.

Sheppard, S.C., M.I. Sheppard, J. Long, B. Sanipelli, and J. Tait. 2006. Runoff phosphorus retention in vegetated field margins on flat landscapes. *Canadian Journal of Soil Science* 86(5): 871-884.

Smyth, R.L., M.C. Watzin, and R.E. Manning. 2007. Defining acceptable levels for ecosystem indicators: integrating ecological understanding and social values. *Environmental Management* (in press).

Soil and Water Conservation Society. 2006. Final report from the Blue Ribbon Panel conducting an external review of the U.S. Department of Agriculture Conservation Effects Assessment Project. Ankeny, Iowa. 24 pp.

Stallins, J.A. 2006. Geomorphology and ecology: unifying themes for complex systems in biogeomorphology. *Geomorphology* 77: 207-216.

Stankey, G.H., R.N. Clark, and B.T. Bormann. 2005. Adaptive management of natural resources: theory, concepts, and management institutions. Gen. Technical Report PNW-GTR-654. Pacific Northwest Research Station, Forest Service, U.S. Department of Agriculture, Portland, Oregon. 73 pp.

Szpir, L.A., G.L. Grabow, D.E. Line, J. Spooner, and D.L. Osmond. 2005. Section 319 national monitoring program projects. Water Quality Group, Biological and Agricultural Engineering Department, North Carolina State University, Raleigh.

Tan P., M. Steinbach, and V. Kumar. 2005. *Introduction to data mining.* Addison Wesley Longman Publishing Inc., Boston, MA.

Townsend, C.R., B.J. Downes, K. Peacock, and C.J. Arbuckle. 2004. Scale and the detection of land-use effects on morphology, vegetation and macroinvertebrate communities of grassland streams. *Freshwater Biology* 49(4): 448-462.

Udawatta, R.P., J.J. Krstansky, G.S. Henderson, and H.E. Garrett. 2002. Agroforestry practices, runoff, and nutrient losses: a paired watershed comparison. *Journal of Environmental Quality* 31(4): 1,214-1,225.

U.S. Army Corps of Engineers. 1999. Rescuing an endangered ecosystem: the plan to restore America's Everglades. http://www.evergladesplan.org/

U.S. Environmental Protection Agency. 2002. National water quality inventory report (305(b) Report): 2000. EPA-841-R-02-001. Washington, D.C.

U.S. Environmental Protection Agency. 2005. Handbook for developing watershed plans to restore and protect our waters. EPA 841-B-05-005. Washington, D.C.

Walters, C.J. 1986. *Adaptive management of renewable resources.* MacMillan, New York, New York. 374 pp.

Walker, B., S. Carpenter, J Anderies, N. Abel, G. Cumming, M. Janssen, L. Lebel, J. Norberg, G.D. Peterson, and R. Pritchard. 2002. Resilience management in social-ecological systems: a working hypothesis for a participatory approach. *Conservation Ecology* 6(1): 14.

Watzin, M.C. and A.W. McIntosh. 1999. Aquatic ecosystems in agricultural landscapes: a review of ecological indicators and achievable ecological outcomes. *Journal of Soil and Water Conservation* 54(4): 636-644.

Watzin, M.C., E.A. Cassell, and D.W. Meals. 2003. Analyzing effects of conservation practices using network modeling. Report to USDA Natural Resources Conservation Service Watershed Science Institute. Washington, D.C.

Wu, J.J., R.M. Adams, C.L. Kling, and K. Tanaka. 2004. From microlevel decisions to landscape changes: an assessment of agricultural conservation policies. *American Journal of Agricultural Economics* 86(1): 26-41.

The Chesapeake Bay experience: Learning about adaptive management the hard way

Thomas W. Simpson
Sarah Weammert

The Chesapeake Bay Program (CBP) is often recognized as a leading cooperative water quality restoration effort that is science-based and uses adaptive management. While CBP officials appreciate the recognition, they are also aware of the limitations and shortcomings of efforts to use adaptive management in water quality restoration efforts. Because the CBP was one of the early cooperative water quality restoration programs, it has been necessary to develop much of the scientific approach used that makes adaptive management essential. Based on experience, however, adaptive management can be difficult to implement in a high-profile program, such as the CBP, with extensive public and political involvement and investment. Despite public and political reaction, CBP officials have begun taking the steps necessary to incorporate adaptive management as a routine part of programmatic decision-making.

Our current philosophy and approach to adaptive management is based upon experiences with the CBP over the last two decades. Clearly, it is important to implement environmental management and controls using the best available science and scientific expertise, based on consensus. Advances in knowledge and experience must be constantly evaluated, but we suggest it be incorporated into the restoration efforts at predetermined intervals. At those times, one can use advances in knowledge and experience to adapt, change, and add management measures. For this to be successful in the public and policy arena, needed adjustments must occur in a transparent, methodical manner.

This relatively simple approach has been challenging to implement, however. Adaptive management can change the outcomes of science-based policy actions, and that can confuse the public and policymakers and cause resistance to implementation of proposed changes. There is no observed problem when adaptive management changes result in greater environmental improvement than projected, as occurred in the CBP watershed with the phosphate detergent ban and biological nutrient removal at wastewater treatment plants. Both efforts yielded greater pollution reduction than anticipated. But this has not been the experience with nonpoint-source pollution controls or best management practices (BMPs). To our knowledge, most, if not all, adaptive management changes to nonpoint-source control measures have reduced the expected impacts of those measures. This can create misperceptions, along with communication and acceptance problems. Policymakers and the public want simple answers, which the CBP provided as annual "progress" based on BMP implementation and model output. Unfortunately, policymakers and citizens do not readily understand the need to adjust the answers as we learn more, particularly when it appears to reduce the impacts of their efforts or make planned efforts more difficult.

Here, we discuss how the CBP became engaged in adaptive management in its watershed policies, strategies, research, and progress reports; lessons learned from conducting adaptive management; and how those lessons have been applied to alter policy and management actions. The discussion provides more detail on our approach to adaptive management and ways to overcome barriers.

Background: The watershed and program

Chesapeake Bay watershed

Chesapeake Bay is the largest and, historically, most productive estuary in the United States. The bay is long and narrow, about 317 kilometers (190 miles) long and 17 to 50 kilometers (10 to 30 miles) wide. The estuary is shallow, with an average depth of 7 meters (22 feet), but it contains a narrow, deep trench up to 50 meters (160 feet) deep that is the drowned valley of the Susquehanna River. What is remarkable about the Chesapeake Bay is the large size of its watershed relative to its small water volume. The watershed covers approximately 160,000 square kilometers (64,000 square miles) (Figure 1). The bay has a watershed area-to-water-volume ratio of about 2400:1, which is six times greater than the estuary with the next largest watershed area-to-water-volume ratio, the Bay of Finland.

The Chesapeake Bay has been in some stage of decline for more than 200 years, with sedimentation, wetland loss, over-harvesting of fisheries, and toxicant and pathogen pollution as principal causes of decline prior to 1950. Congress authorized a major research effort in the mid-1970s to determine the cause of accelerated decline in habitat and living resources in the bay. When this research was completed, it became evident that nutrient enrichment was the principal cause of the more recent, systemic decline (U.S. Environmental Protection Agency, 1999).

Nutrient enrichment, or eutrophication, has caused excessive algal growth, which can result in areas of low or no oxygen in most deep waters and some shallow creeks and rivers from May through September. This effectively eliminates the cooler, deeper waters as warm weather habitat for finfish and shellfish and makes survival of benthic organisms difficult. In shallow tidal rivers and creeks, low oxygen is responsible for many reported fish kills, particularly during spring. The excessive algal concentrations also impair clarity in shallow water and, in tandem with sediments, are responsible for turbidity that has resulted in the loss of most of the underwater grasses in the tidal shallows [0.5- to 2.0-meter (1.7- to 6.6-foot) depths]. These grasses are critical habitat for many species of finfish and crabs; they also serve as filters to improve water quality and clarity and buffer shorelines from wave action (Koroncia et al., 2003).

Estimated nitrogen and phosphorus loads are more than six times presettlement loads. All sources of human activity contribute to these loads, but agriculture is a major source of nutrients in the watershed. Reducing the impact of nutrients on Chesapeake Bay water quality cannot be achieved by point-source reductions alone, but must include extensive reductions from nonpoint sources, particularly agriculture. Figure 2 presents a model-based estimate of the relative contribution of different sources to nutrient loads in 2003. Agriculture is a major contributor of nutrients to the bay and a focal point of all state tributary strategies.

Chesapeake Bay agreements and program

Declines in the extent of underwater grasses and expansion of anoxic zones during the 1960s and 1970s led to a congressionally mandated study to determine causes of the decline. When this study culminated in 1983, the states of Virginia, Maryland, and Pennsylvania, along with the District of Columbia and the federal government, signed the first Chesapeake Bay Agreement, committing in very general terms to work together to reduce nutrient pollution in the bay. The agreement created the CBP and directed officials with that program to take the lead in managing activities aimed at restoring the bay. The CBP is a regional partnership that brings together members of various state, federal, academic, and watershed organizations to build programs and adopt policies that support bay restoration.

Since 1983, two additional agreements and one major amendment have been signed that guide bay restoration and management policies and activities. In 1987, the second Chesapeake Bay Agreement established the CBP's goal to reduce the amount of nutrients, primarily nitrogen and phosphorous, that enter the Bay by 40 percent from 1985 levels by the year 2000 (U.S. Environmental Protection Agency, 1987). The agreement continued cooperative efforts to restore and protect the bay, further committed partners to specific restoration actions, and declared that implementation of those commitments would be reviewed annually and additional commitments developed as needed. The new agreement also contained goals and priority commitments for living

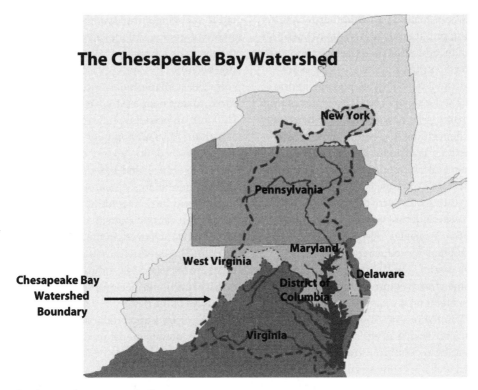

Figure 1. The Chesapeake Bay watershed.

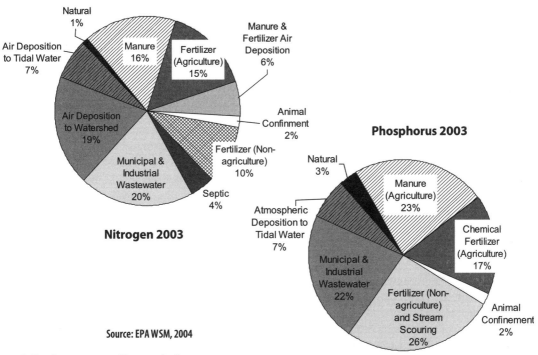

Figure 2. Nutrient sources to Chesapeake Bay.

resources; water quality; population growth and development; public access; governance; and public information, education, and participation, but the "40 percent" nutrient reduction goal dominated further activities. The "40 percent" reduction was applied only to controllable sources from within signatory jurisdictions and did not initially include air deposition. As a result, the goal was actually a reduction of about 23 percent of the total nitrogen and phosphorus loads to the bay.

In 1992, the 1987 Chesapeake Bay Agreement was amended to require jurisdictions to develop subbasin-specific nutrient reduction strategies, called tributary strategies, designed as river-specific cleanup plans for reducing nutrient levels in each subbasin and the bay itself by 40 percent. The amendment also required tributary-specific strategies to maintain the capped loading goals beyond 2000 in the face of growth and development. Maryland used local agricultural tributary teams to develop agricultural strategies based on estimated levels and impacts of BMP implementation. A "technical options team" developed BMP efficiencies based on a 1990 Chesapeake Bay Program BMP report, the scientific literature, and expert consensus. This was used by the local tributary teams in each of Maryland's 10 subbasins to develop strategies appropriate for, and implementable in, their basin that were estimated, based on BMP efficiencies and model projections, to achieve the needed 40 percent reduction if rigorously implemented. No post-2000 capped load management strategies were developed. The BMP definitions and efficiencies were adapted for basinwide use for all nonpoint sources in 1995.

In June 2000, bay program partners adopted Chesapeake 2000, an agreement intended to guide restoration activities throughout the bay watershed through 2010. In addition to identifying key measures necessary to restore the bay, Chesapeake 2000 provided the opportunity for Delaware, New York, and West Virginia to become more involved in the bay program partnership by working to reduce nutrients and sediment flowing into rivers from their jurisdictions. In 2000, New York and Delaware committed to the Chesapeake 2000 water quality goals; West Virginia did so in 2002. The agreement committed the CBP to reducing total nutrient loads by nearly 50 percent or about twice the "40 percent" controllable load agreed upon in 1987. The 2000 agreement commit-

ted to reducing nutrient and sediment loads sufficiently to remove all related water quality impairments in the bay, with a goal of accomplishing this by 2010.

In 2003, CBP partners agreed to reduce nutrient loads so that no more than 80 million kilograms (175 million pounds) of nitrogen and 6 million kilograms (12.8 million pounds) of phosphorus would be delivered to the bay. Partners also agreed to reduce land-based sediment loads so that no more than 3.77 metric tons (4.15 million tons) would be delivered to the bay. These were considered capped loads that had to be maintained, once achieved, in the face of subsequent growth. These reductions in nutrients and sediment are expected to produce the improved water quality conditions necessary to support the living resources of the bay.

At the agreed-upon reductions, the Chesapeake Bay water quality model predicts the existence of a bay similar to that in the 1950s. Water quality conditions necessary to protect aquatic living resources will be met in 96 percent of the bay within designated use requirements, and the remaining 4 percent will fall shy of fully achieving water quality conditions for only four months a year.

Between 1985 and 2002, annual phosphorus and nitrogen loads delivered to the bay from the entire watershed (New York, Pennsylvania, Maryland, Delaware, District of Columbia, West Virginia, and Virginia) were estimated to decline by 7.64 million pounds and 59.81 million pounds, respectively, based on watershed model projections. Monitoring data at river input stations showed significant reductions on most rivers, but not as much as projected by modeling. Tidal monitoring in the bay's mainstream showed no significant change. The reductions obtained between 1985 and 2002 included offsetting significant increases due to population growth. Maintaining reduced nutrient and sediment levels into the future will be a challenge because of expected growth in human population and shifts in animal manure nutrients and the land available for application of those nutrients.

In 2003 tributary strategies were completed by jurisdictions to implement the agreed-upon reductions. Nutrient loads were allocated to nine major tributary basins that were further divided into 16 major tributary basins by jurisdiction, then into 37 jurisdictionally defined tributary strategy subbasins. As in 1993, the 2003 tributary strategy process began with a strong emphasis on stakeholder involvement. But these goals were more challenging, and many of the more easily implemented practices were in place. As a result stakeholders could not agree on an implementation strategy that model results indicated could achieve proposed loading goals. Jurisdictions then developed strategies that were estimated to achieve the loading goals based on BMP efficiency estimates and watershed model output. Those strategies required near complete implementation of many existing practices and incorporated numerous new BMPs. Nearly all jurisdictions also required major additional nutrient reductions from point-source discharges. The nonpoint-source strategies focused heavily on agricultural BMPs because they are more cost effective than urban BMPs. Table 1 lists the agricultural BMPs used in the tributary strategies. BMP efficiency estimates and the model underwent minor revisions in 1997 and 2000. Some new practices were added prior to the new tributary strategies, but these did not change estimated outcomes substantially.

During the 1990s, the CBP began using model estimates of reductions in nutrient loads as surrogates for progress in water quality improvement. These were comparatively easy to generate; provided simple, understandable progress estimates; and were guaranteed to show progress proportional to implementation because annual loads, regardless of the actual year, were based on average load estimates using 1985-1994 hydrologic conditions. Average hydrology was used to allow trend analysis over highly variable climatic conditions and was thought to allow estimation of progress delayed in occurring because of lag times. By the late 1990s, it became apparent that modeled progress was overestimating actual progress and that much of this overestimation was related to assumptions about and reporting and application of BMPs used in the watershed model. As mentioned, minor revisions in the BMP approach were made in 1997 and 2000, but it was not until a 2003 CBP Scientific and Technical Advisory Committee Forum and subsequent white paper (Simpson et. al., 2003) clearly stated the issues and concerns that substantial actions were taken to revise the approach. This resulted in substantial changes to BMP efficiency estimates during 2003 that are discussed below.

The white paper was one of several factors that led to media, agency, and congressional investigations and reviews of CBP nutrient reduction activities. Those investigations and reviews helped drive the changes that resulted in a new approach to progress indicators in 2006. Most of the reports concluded that modeled information was being overused to estimate progress in water quality and living resource restoration. This was not intended to be misleading, but these conclusions developed because of inadequate knowledge regarding BMP effectiveness, implementation, operation and maintenance, and agency tracking and reporting procedures. The BMP revisions initiated in 1997, but not undertaken until 2003, underscored the need for the new approach to progress estimation released in 2006 and represented successful use of adaptive management in a policy setting.

Lessons learned

The Scientific and Technical Advisory Committee white paper identified the technical issues that are the apparent cause of the overly optimistic BMP reduction estimates. First, lag times are real and may delay actual water quality improvements for years to decades, but hydrologic lag times are only one source of delay. It is apparent that delays in planning, implementation, and practice maturity are also important factors that must be considered when estimating when actual practice effects will be seen. Hydrologic lag times are well documented. Dissolved nitrogen associated with groundwater may have a transport time of years to decades, with a median time of about 10 years (Lindsey et al., 2003). Nutrients associated with sediment can have much longer transport times (several decades) in the watershed because of their storage in soil and stream corridors, both of which are greatly influenced by annual rainfall. Additionally, the location of the source area in the watershed will influence the lag time between implementation lag times and improvement in bay water quality. Planning, implementation, and practice maturity lag times may be easier to estimate than hydrologic lag times, but are rarely considered. It was apparent, however, that other factors besides lag times were influencing the difference between estimated and observed practice impacts.

Table 1. Tributary strategy agricultural BMPs, 2005.

BMPs with CBP or jurisdictional (*) efficiencies	BMPs requiring peer review
Riparian forest buffers	Continuous no-till
Riparian grass buffers	Dairy precision feeding and forage management
Wetland restoration*	Swine phytase
Land retirement	Ammonia emission reductions
Tree planting	Precision agriculture
Conservation tillage	Precision grazing
Carbon sequestration/alternative crops*	Water control structures
Poultry phytase	Stream restoration
Poultry litter transport	
Nutrient management	
Enhanced nutrient management*	
Conservation plans/SCWQP	
Cover crops (early- and late-planting)*	
Small grain enhancement (early- and late-planting)*	
Off-stream watering w/ fencing	
Off-stream watering w/o fencing	
Off-stream watering, fencing and rotational grazing*	
Animal waste management systems: livestock	
Barnyard runoff control/loafing lot management	
Animal waste management systems: poultry	

Source: Perkinson, 2004.

The other primary factors that caused overly optimistic assumptions about progress were BMP efficiency and application assumptions. Probable sources of error in estimating BMP impacts include limited data and/or field observation; research and plot-scale reduction efficiencies applied to watershed-scale implementation; extreme spatial variability in soils, hydrology, and management; assuming proper implementa-tion, operation and maintenance, function, and replacement; and finally, optimistically reported implementation rates. These all appear to result in greater pollution control credits for practices than what is likely to occur.

As a result, in 2003, selected BMPs were evaluated and their efficiencies revised to better reflect current research and knowledge; the goal was to provide more realistic estimates of expected pollution reduction levels. Nutrient management plan efficiency estimates, for example, were reduced from 30 percent to 24 percent, although the initial technical proposal would have reduced them to 18 percent. Conservation plan and cover crop (applied to conservation tillage) efficiency and implementation assumptions were also reduced. This affected watershed model output by reducing apparent "progress" and making it more difficult to develop tributary strategies that would achieve loading goals. For both progress runs and strategy scenarios, the changes resulted in substantially lower modeled phosphorus reductions and somewhat lower nitrogen reductions.

Policymakers and program managers reacted to the proposed changes much differently than did people in the scientific community. Individuals in the scientific and technical communities supported the changes. They agreed with the evidence that suggested BMP impacts were overestimated and that changes in reduced modeled "progress" would bring model results closer to expected water quality. There is no "good time" to change BMP assumptions, but they must be made at recurring intervals. Changes should not be made continuously because policy cannot accommodate frequent adjustments.

Agency managers responsible for explaining the apparent reduced progress to their stakeholders and policymakers were put in the difficult situation of having to explain the slower progress and increased difficulty in achieving already challenging loading goals. Stakeholders and policymakers were initially confused, and that detracted from the accelerated implementation that was needed. It was apparent that the changes were needed and did in fact improve estimated impacts and, ultimately, CBP credibility. But neither managers nor scientists could afford to lose stakeholder or policymaker support. With time, most people have recognized the need for and value of the changes, but initial acceptance was slow in coming.

Our experience has helped to identify approaches for conservatively predicting conservation effects and using adaptive management in water quality restoration programs to avoid some of the difficulties we encountered. Refining efficiencies is an iterative, adaptive process that must continue while research identifying and adding new BMPs is being conducted. It is essential to be conservative in setting and revising assigned efficiencies. From a practical standpoint, it is always easier to increase practice efficiencies than to reduce them from a policy and public acceptance perspective. There are no known examples where BMP efficiencies underestimate pollution reduction impacts, reinforcing the need to use conservative efficiency estimates.

Research- and demonstration-site-derived efficiencies for watershed-scale implementation efforts do not reflect the spatial variability of an entire watershed. Efforts should be made to assure that reported implementation is close to actual and to determine if farmer implementation and operation is as rigorous as specified in the practice. Model output and monitoring data must be consistent and used appropriately. Implementation progress, water quality restoration progress, and habitat/living resource and restoration progress should be reported separately, using different data sources and approaches. This will more accurately convey the health of the ecosystem to stakeholders and is critical to maintaining the long-term momentum of any restoration effort. Better research on demonstration and monitoring of individual BMPs, conservation systems, and small watershed conservation effects will increase confidence in BMP effectiveness. Finally, managers, policymakers, and involved citizens must be made aware of potential implications of adaptive science and understand why an adaptive approach is essential.

Applying the lessons learned

CBP officials have initiated changes in BMP applications in the watershed model and in use of model output to estimate progress as a means of improving management procedures based upon the lessons learned. Operational assumptions and reporting issues related to BMP performance during the calibration of the new phase of the Chesapeake Bay watershed model are being addressed. Jurisdictions are now working to ensure the accuracy of their reported BMP implementation levels. CBP staff members are also working to assure consistency between and among jurisdictional reporting methods.

A project has also been funded to develop science-based definitions and efficiencies for new BMPs and to review existing BMP definitions and

Agricultural Pollution Controls

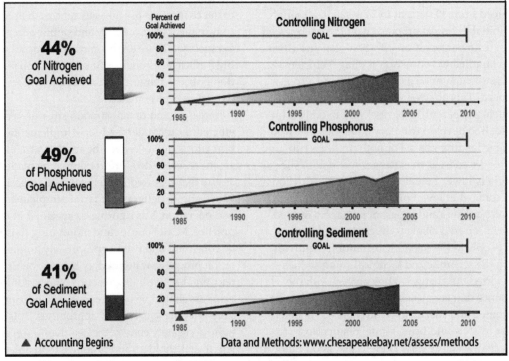

Source: Chesapeake Bay Program (www.chesapeakebay.net/assess/methods)

Figure3. Implementation progress with agricultural tributary strategies (U.S. Environmental Protection Agency, 2006b).

efficiencies. This project will allow these practices to be incorporated into the new phase of the watershed model for calibration and subsequent progress runs of tributary strategies. The project is designed to assure that the best available information for a specific BMP will be used to draft or refine each BMP definition and efficiency. As a result of more accurate calibration and model outputs, jurisdictions will be able to adjust their tributary strategies to more credibly and efficiently achieve nutrient and sediment loading goals. Because the involved parties now understand the need for adaptive management and better data, additional research is being funded or prioritized in request for proposals by the CBP and other funding sources for projects that enhance the science base for BMP efficiency and implementation estimates.

The CBP has adopted a new approach to progress reporting that separates the different components of progress (U.S. Environmental Protection Agency, 2006a). Implementation progress is used to estimate progress toward use of practices

and control measures at the levels contained in jurisdictional tributary strategies. Figure 3 shows implementation progress with the agricultural components of tributary strategies summarized throughout the bay's watershed. Implementation progress is separated from water quality restoration progress, which is shown as trends for nitrogen at river input stations in Figure 4. This does show significant declines in nitrogen loads in most rivers, but those declines are less than needed to remove water quality impairments. It should be noted that tidal mainstem monitoring has yet to show significant nutrient reductions. But river input monitoring clearly suggests reductions in loads from the watershed in the face of population growth and development.

Living resources restoration related to reduced nutrient loads is currently being estimated by improvements in key habitat controlling parameters, such as dissolved oxygen, as illustrated in figure 5. While the trend is slightly positive, it is not yet significant. The ultimate living resource restoration indicators are healthy, diverse stocks

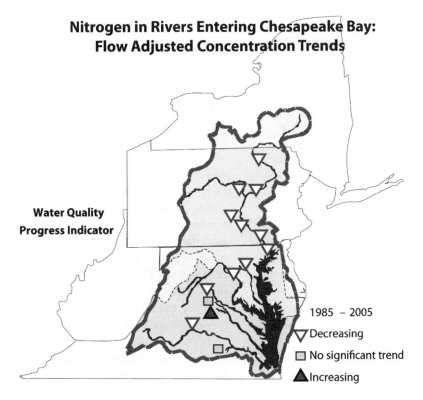

Figure 4. Trends in nitrogen loads entering the Chesapeake Bay at river input stations.

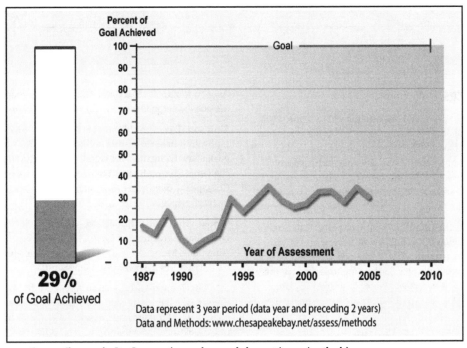

Figure 5. Trends in the annual rolling average of dissolved oxygen standard attainment in the Chesapeake Bay (U.S. Environmental Protection Agency, 2006c).

of fish and shellfish, dependent upon deep waters, with adequate oxygen and grasses in clear, shallow water.

Summary and conclusions

Adaptive management allows implementation, management, and policy to move forward as knowledge improves, but changing the science in a policy and management setting is challenging and difficult. Adaptive management has scientific, communication, and policy implications.

The CBP work throughout the 1990s to use BMP efficiencies was pioneering and offers lessons and direction to others trying to represent conservation effects in a realistic fashion. It is important to move forward using the best available science while being conservative enough to avoid overestimation of impacts that create unrealistic expectations.

Policymakers and managers must better understand the use of adaptive management so they will accept changes as they are proposed. It is also important to make changes in conservation effect

estimates at planned intervals. It is difficult to implement policy and management programs if the impact estimation science is constantly changing, but that science must be incorporated over time. Experience in the Chesapeake watershed suggests three- to five-year intervals are manageable from a policy perspective and short enough that major changes in knowledge regarding the science of conservation effects is unlikely. It is also essential that adequate research be conducted on conservation effects at field and watershed scales; spatial and management variability; and enhanced understanding of factors influencing adoption, implementation, operation, and maintenance of practices so that uncertainty in estimates of conservation effects can be reduced. Adaptive management is essential when applying science to policy, but use of adaptive management presents challenges at the interface between science and policy. Those challenges can be diminished, however, through expanded knowledge of the real conservation effects of practices, systems, and programs.

References

Koroncai, R., L. Linker, J. Sweeney, and R. Batiuk. 2003. Setting and allocating the Chesapeake Bay Basin nutrient and sediment loads: The collaborative process, technical tools and innovative approaches. Chesapeake Bay Program Office, U.S. Environmental Protection Agency, Annapolis, Maryland. <http://www.chesapeakebay.net/caploads.htm>

Lindsey, B.D, S.W. Phillips, C.A. Donnelly, G.K. Speiran, L.N. Plummer, J.K. Bohlke, M.J. Focazio, W.C. Burton, and E. Busenberg. 2003. Residence times and nitrate transport in ground water discharging to streams in the Chesapeake Bay watershed. Water Resources Investigations Report 03-4035. U.S. Geological Survey, Baltimore, Maryland. 201 pp.

Perkinson, R. 2004. Agricultural BMP descriptions as defined for the Chesapeake Bay Program watershed model. Chesapeake Bay Program, Agricultural Nutrient Reduction Workgroup, U.S. Environmental Protection

Agency, Annapolis, Maryland. <http://www.chesapeakebay.net/pubs/waterqualitycriteria/doc-Ag_BMP_Defns.pdf>

Simpson, T.W., C.A. Musgrove, and R.F. Forcak. 2004. Innovation in agricultural conservation for the Chesapeake Bay: Evaluation progress and assessing future challenges. Scientific and Technical Advisory Committee, Chesapeake Bay Program, Annapolis, Maryland. <http://www.chesapeake.org/stac/stacpubs.html>

U.S. Environmental Protection Agency. 1983. The Chesapeake Bay Agreement of 1983. Chesapeake Bay Program, Annapolis, Maryland <http://www.chesapeakebay.net/pubs/1983ChesapeakeBayAgreement.pdf>

U.S. Environmental Protection Agency. 1987. The 1987 Chesapeake Bay Agreement. Chesapeake Bay Program, Annapolis, Maryland. <http://www.chesapeakebay.net/pubs/1987ChesapeakeBayAgreement.pdf>

U.S. Environmental Protection Agency. 1999. Chesapeake Bay Program, Annapolis, Maryland. <http://www.chesapeakebay.net/agreement.htm>

U.S. Environmental Protection Agency. 2004. Estimated nitrogen and phosphorus loads by source. Chesapeake Bay Program, Annapolis, Maryland.

U.S. Environmental Protection Agency. 2006a. Long term plan for communicating the state of the bay and the state of the bay restoration (version 9/19/06). Chesapeake Bay Program, Indicator Workgroup, Annapolis, Maryland. <http://www.chesapeakebay.net/pubs/subcommittee/ irw/LongTermPlanImplementingCBPIndicatorsFramework091906.doc>

U.S. Environmental Protection Agency. 2006b. Chesapeake Bay 2005 health and restoration assessment; part one: Ecosystem health. Chesapeake Bay Program, Annapolis, Maryland. <http://www.chesapeakebay.net/assess/ health_report_final_033106.pdf>

U.S. Environmental Protection Agency. 2006c. Chesapeake Bay 2005 health and restoration assessment; part two: Restoration efforts. Chesapeake Bay Program, Annapolis, Maryland. <http://www.chesapeakebay.net/ assess/restoration_report_final_033106_web.pdf>

Part 5

Research information needs: Constituency perspectives

A local stakeholder perspective

Steve John

As a local stakeholder, I have worn several hats over a span of about 30 years: Interested citizen; Decatur, Illinois city council member (1987-1995); co-founder and active member of the Upper Sangamon River Watershed Committee; environmental planning consultant; and now executive director of the Agricultural Watershed Institute, a nonprofit research institute established in 2003. AWI's mission is to conduct research and educational programs on practices and policies that improve water quality, maintain or restore ecosystem health, and conserve and manage land and water resources in agricultural watersheds. As a councilman, it was my job to represent the interests of a municipal water utility impacted by land use and agricultural practices in its watershed. As former co-chair of the watershed committee and now with AWI, my challenge was and is broader: To help identify and study solutions that all affected stakeholders can support.

Sediment and nitrates in Lake Decatur

The Lake Decatur watershed covers about 2,313 square kilometers (925 square miles). About 87 percent of this area is used for production of two crops, corn and soybeans. Much of the prime farmland in the watershed is on nearly level

or gently sloping uplands with subsurface tile drainage.

1922: Water for grain processing and domestic use. In Decatur's early history, water for domestic and industrial use was drawn from wells and later from the Sangamon River. The dam that formed today's Lake Decatur was constructed in 1921-1922. Local business leader A. E. Staley was a key figure in the decision to build the dam. His company, then known as the Staley Starch Works, needed more water to process grain.

1920–1970: Change in agricultural land use. By 1922, corn was already the principal local crop, although hay, wheat, and oats combined almost equaled corn acreage in the watershed. Soybean acreage increased from near zero in 1920 to surpass hay, wheat, and oats by 1940. By the 1970s, corn and soybean acreage was nearly equal; hay crops and small grains had almost disappeared from the agricultural landscape.

1940s: Soil conservation efforts begin. Decatur began to implement "source-water protection" long before that term was written into the 1996 Safe Drinking Water Act. Soon after construction of the dam, sediment problems became all too apparent in Lake Decatur. In the early 1940s, soil conservationists employed by the city were actively involved in promoting soil conservation and helping to form soil and water conservation districts in the counties of the Upper Sangamon watershed.

1967: One elevated nitrate sample. A water sample drawn from Lake Decatur in 1967 was found to have a nitrate-nitrogen concentration of 12 milligrams per liter (12 parts per million). Up to that time, sampling was done only sporadically, but no sample had ever tested above 10 milligrams per liter. Illinois statewide agricultural statistics show that from near zero in 1950 annual application of inorganic nitrogen fertilizer increased to a little more than 100,000 tons by 1960 and then climbed rapidly to about 600,000 tons by 1967. After occurrence of this first elevated nitrate concentration, regular sampling was started. The 10-milligram-per-liter level was not exceeded again until 1980. By that year, average annual application of inorganic nitrogen fertilizer in Illinois was about one million tons.

1980–present: Annual nitrate excursions. Since 1980, the 10-milligrams-per-liter nitrate drinking water standard has been exceeded in Lake Decatur in most years for a period ranging from weeks to months, with the highest concentrations occurring in the spring.

1985–present: City-farm cooperation. In 1985, a Sedimentation Control Committee appointed by Decatur's mayor presented recommendations for pilot studies and dredging portions of the lake. The committee report also said: "The City must aggressively encourage soil conservation practices in the Lake Decatur Watershed and, in dialogue with Macon County Soil and Water Conservation District, should determine a level of long-term financial funding for that purpose." The urban-rural partnership formed in response to that recommendation endures today.

1992–2002: Monitoring, studying options, and constructing nitrate treatment facilities. After a decade of annual excursions above the nitrate drinking water standard, the Illinois Environmental Protection Agency (IEPA) in 1992 required the city to sign a letter of commitment to select and implement an alternative to achieve compliance with the standard. A draft letter of commitment, prepared by IEPA, said that the city would either install a nitrate removal treatment process or find an alternative low-nitrate water source. The city requested, and IEPA agreed, to add watershed management to the list of potential alternatives.

After signing the letter of commitment, the city contracted with the Illinois State Water Survey to monitor nitrates at several stations on the Sangamon River and its tributaries. Water quality monitoring, studies of watershed and engineered alternatives, and public outreach continued from 1992 to 1998. A survey of agricultural producers suggested that average nitrogen application rates were significantly higher than University of Illinois recommendations. Modeling of water quality response to projected voluntary adoption of best management practices (BMPs) suggested that a ramped-up program of education and outreach, technical assistance, and BMP cost-share could reduce, but not eliminate, excursions above the nitrate standard. The city proposed a compliance plan that included an expanded watershed management program to reduce peak nitrate concentrations in the lake, combined with an engineered approach with a design capacity reflecting the projected benefits of the watershed management effort. In effect, the city proposed adaptive implementation of a voluntary watershed strategy,

with continued monitoring to confirm whether it was achieving the projected reduction in nitrate concentrations.

In the end, IEPA required the city to install nitrate treatment facilities – an ion-exchange system – designed for the historic peak nitrate concentration in the lake. This implicitly sent the message that, while the city was welcome to continue promoting nitrogen BMPs, the benefits were too uncertain to be included in a formal plan for drinking water compliance. Design of the ion-exchange facility started in 1999. Construction was completed and the system placed into service in 2002 at a capital cost of $7.6 million.

1993–present: Erosion reduction and dredging program. While the city was studying, designing, and constructing its nitrate removal system, it was also continuing its financial support of conservation district staff and locally designed cost-share programs aimed at controlling soil erosion and reducing sedimentation in the lake. The city began a phased dredging program to restore reservoir capacity and maintain the lake's aesthetic and recreational value. In 1993-1994, one portion of the lake was dredged, with about 1,300 acre-feet of sediment removed and pumped to a large sediment basin to dry. Dredging now underway and planned future phases will remove an additional 4,000 acre-feet over the next decade. AWI is working with the city and other local stakeholders to evaluate options for large-scale beneficial use of dredged sediment. Small-scale demonstrations have been conducted of application of dried sediment to eroded agricultural land and blending of sediment with composted plant material to produce a topsoil for horticultural use. Additional demonstration projects are planned.

Science and Lake Decatur watershed management today

Decatur officials responsible for source-water protection and conservation districts working in the watershed make day-to-day decisions about getting the most bang for the buck by prioritizing allocation of personnel resources and cost-share dollars. Conservation district and Natural Resources Conservation Service (NRCS) employees keep up with research on new and improved practices. Decatur is the location for the Farm Progress Show held in Illinois every other year.

Macon County Conservation District has worked closely with university researchers, agricultural producers, contractors, and manufacturers to make the Farm Progress Show site a showplace for soil erosion reduction and drainage water management practices.

The city's main focus is the never-ending effort to reduce soil erosion and sedimentation. Clearly benefits of erosion reduction accrue to landowners interested in the long-term productivity of the soil. From the city's perspective, the bottom-line question is this: Does a dollar of city taxpayer or ratepayer money spent on watershed programs generate more than a dollar's worth of benefits in maintaining reservoir capacity, protecting an important recreational amenity, and reducing the need for future dredging of the lake? Scientific research at the field or landscape scale can contribute to improving the benefit-cost ratio of public investment in erosion reduction and other source-water protection activities.

Having expended capital funds for the ion-exchange treatment system, the city has less reason to focus on nitrate reduction, but continues to support this as a secondary priority for its watershed activity. The annual cost of operating the ion-exchange facility is about $1 million. If successful, nutrient management efforts may reduce operating costs. In addition, there are grounds for concern that rising corn prices, driven in no small part by the push for corn-derived ethanol, may result in increased nutrient loads and concentrations in surface water. Without serious efforts to mitigate nitrogen losses from tile-drained cropland, the city may confront a need to add nitrate treatment capacity in the future.

Given its mission to address water quality issues related to agriculture land use, AWI is interested in research on ways to mitigate not only local impacts – nitrates and sediment in Lake Decatur – but also downstream impacts, including hypoxia in the Gulf of Mexico. AWI is committed to seeking and evaluating win-win solutions that benefit the environment without adversely affecting net farm income. The Upper Sangamon watershed provides a laboratory without walls for adaptive implementation of nutrient management practices to advance these goals. The data record contains nearly 40 years of nitrate sampling in Lake Decatur and 15 years of river and tributary monitoring. The agricultural landscape provides

sites suitable for testing a variety of fertilizer management, subsurface and surface drainage management, riparian and wetland systems, alternative cropping systems, and market mechanisms, such as risk reduction and water quality trading concepts.

AWI performs collaborative research and demonstration on watershed management strategies. The Upper Sangamon River Targeted Watershed Project, with grant funding from the U.S. Environmental Protection Agency, engages conservation districts, the fertilizer and drainage industries, University of Illinois researchers, and cooperating producers and landowners in several studies of nitrogen and phosphorus fertilizer management practices and subsurface bioreactors. This project includes paired subwatersheds in which the Illinois State Water Survey is monitoring flow and nutrients.

On another current project, funded in part by the Illinois Clean Energy Community Foundation, AWI is leading a learning group to assess economic feasibility and environmental benefits of biomass energy. Participants in this cross-sector learning group come from agriculture, conservation organizations, government agencies, industry and electrical utilities, and the University of Illinois.

Lessons for science-based watershed management

Based on my long involvement with the Lake Decatur watershed and recent experience with AWI, I offer these thoughts on ways to strengthen the science base of watershed management. But let me attach one disclaimer: These are my personal observations and suggestions. They have not been vetted by the AWI board of directors or local stakeholders in the watershed.

1. *There is a need for land use and agricultural statistics reported on a river basin or major watershed basis.* While acknowledging the challenges, I would suggest that to study the effects of land use and agricultural practices on water quality (and quantity) at a watershed scale we must have data reflecting what is happening in the watershed. Statistical manipulation of county or regional data to approximate watershed data inevitably weakens the analysis of correlations and

causation. For retrospective studies correlating land use and water quality, we have to do the best we can with what is available. Going forward, we should be able to apply advances in geographic information systems to provide spatially referenced data in formats amenable to testing hypotheses about water quality impacts of agriculture and other land uses.

2. *There is a need to develop and test protocols for studies on the efficacy of voluntary watershed management strategies.* An ongoing debate plays out in local total maximum daily load (TMDL) public meetings and state or federal policy hearings between advocates of voluntary and regulatory approaches to nonpoint-source pollution abatement. I would suggest that the efficacy of voluntary approaches (including strategies incorporating economic incentives and disincentives) is a researchable question. Three broad questions need to be answered:

- Were the proposed voluntary actions implemented?
- What were the economic outcomes?
- What were the environmental outcomes?

The details will, of course, vary with circumstances. But general methods and good case studies of this type of research would be very helpful. If proponents of voluntary strategies (and I count myself among them) are serious about meeting environmental objectives, they should be willing to support systematic measurement to assess whether recommended practices were adopted and achieved their objectives.

3. *There is a need for more long-term studies on paired watersheds to examine a range of practices, cropping and drainage systems, and other variables.* To a limited extent, this may be accomplished on land owned or controlled by universities or other research organizations. But small watersheds with suitable physical characteristics are likely to include numerous parcels with multiple private owners and producers. To establish long-term studies at a landscape or watershed scale, ways are needed to include landowners and producers as active participants in the research enterprise. Their participation needs to be appropriately valued and compensated, including protection against lost income or adverse

effects on the land resource resulting directly or indirectly from the research. This calls for formal agreements between research organizations and cooperating landowners and producers that go beyond the usual small incentive payments of most short-term, field-scale studies.

4. *There is a need for refined estimates and measurement of the water quality benefits that can be realistically achieved at a watershed scale through adoption of specific practices or sets of practices.* To some extent, this is a matter of calibrating water quality models at the watershed scale. This would make for more credible TMDL studies and implementation plans. I also have in mind the kind of meta-analysis of case studies that would be of practical value to participants in local watershed planning committees or larger efforts, such as the Upper Mississippi River Subbasin Hypoxia Nutrient Committee formed under the umbrella of the Mississippi River/Gulf of Mexico Nutrient Task Force. I expect that such an analysis would underscore the need to think beyond our current menu of BMPs.

5. *There is a need – and an opportunity – to develop new agricultural systems to improve water quality and meet other critical goals. Scientists, policymakers, agribusiness, and agricultural producers all have an important role in this ambitious undertaking.* I would suggest that continuous improvement in crop and livestock management practices is necessary, but often not sufficient, to achieve water quality goals. As the exploration of ecosystem services illustrates, farsighted scientists, policymakers, conservation practitioners, and agricultural producers are beginning to formulate a vision for a new agriculture that can meet society's need for food, fiber, and energy; provide increased and diversified farm income; enhance water and air quality and habitat; and protect soil and water resources. AWI is involved in one such cooperative effort as a member of the Green Lands, Blue Water Consortium. Consortium members are land grant universities and nongovernmental organizations. Its mission is to "support development of and transition to a new generation of agricultural systems in the Mississippi River Basin [that] integrate more perennial plants and

continuous living cover into the agricultural landscape." The roadmap for achieving the consortium's landscape vision emphasizes enterprise development, with scientists and stakeholders from various economic sectors working together to expand uses and markets for perennials and annual cover crops. AWI is especially interested in the economic and environmental benefits of perennial crops grown for biomass energy. A working hypothesis of its biomass energy learning group is that cropping systems incorporating herbaceous or woody energy crops, along with annual row crops, can enhance habitat and water quality and provide renewable energy feedstocks. The linked imperatives of reducing greenhouse gases and achieving energy self-reliance are seen as a powerful driver for fundamental changes in agriculture in the coming years. Scientists, including economists and social scientists, will be instrumental in developing prescriptions for what to grow, where and how to grow it in order to optimize production and ecosystem benefits, and how to convert biomass into heat and power or liquid fuels. Policymakers, conservation practitioners, and, most importantly, farmers, will be instrumental in making this happen.

A state agency perspective

Todd Ambs

As I looked at the key themes of this workshop, I especially liked the focus on "realistic expectations." My world is a world of high hopes and low expectations.

The Water Division at the Wisconsin Department of Natural Resources (DNR) has the greatest scope of any such division in the nation. The division is the only one in the nation with Clean Water Act and Safe Drinking Water Act duties in the same place as fisheries management. The structure is designed to follow an integrated resource management model promoted by Aldo Leopold.

We have many world-class water resources to protect in Wisconsin—a land of 15,000 lakes, 140,000 kilometers (84,000 miles) of rivers, 2.1 million hectares (5.3 million acres) of wetlands, 1,833 kilometers (1,100 miles) of Great Lakes shoreline and enough groundwater that if it were laid evenly across the surface of the state it would be 30.3 meters (100 feet) deep.

The state's number one water quality problem is nonpoint-source or polluted runoff. It is a persistent problem for 40 percent of rivers and 90 percent of lakes in Wisconsin. Although the state has many forms of polluted runoff, one of the most vexing is due to improper manure management. From May 2004 through July 2005, 52 signifi-

cant runoff events from manure spreading were recorded. There may have been even more unrecorded events. Those events killed literally tens of thousands of fish in numerous streams and contaminated a number of private drinking water wells in several areas of the state.

Despite limited resources and staffing constraints, DNR officials are doing their best to address these issues. New rules for concentrated animal feeding operations (CAFO) have been proposed. Some key features of those new rules include the following:

1. Prohibit applying liquid manure on frozen or snow-covered ground unless it's injected or immediately incorporated into the soil or unless it's an emergency outside the operator's control. Solid manure spreading would be prohibited on frozen or snow-covered ground during February and March unless it's immediately incorporated.

2. Require six months of liquid manure storage, with some exceptions. Up to 80 percent of DNR-permitted livestock producers already have, or plan to have, six months of storage for liquid manure. Illinois, Michigan, Ohio, and Indiana already require at least this much storage.

3. Require that manure spread on land be set back from private and public drinking water wells and from sinkholes and fractured bedrock. Additional restrictions would apply to manure and process wastewater spread on areas with shallow soils.

4. Require farms to implement nutrient management plans based on applying the right amount of phosphorus, a nutrient, which if it enters lakes and rivers, reduces water quality and fuels algae growth.

Wisconsin also has proposed new rules to regulate the use of pesticides containing alachlor, a known carcinogen.

In late 2002, the state adopted the most comprehensive nonpoint rules in the nation. Those rules are now being implemented. Good progress has been made, but work has been slowed on the agricultural side by a cost-share provision. Under state law, farms other than CAFO's must be offered at least 70 percent cost sharing to install practices to correct a water quality problem. The lack of state cost-share dollars has hampered the ability to require corrective measures in many areas.

Some reactions to the state's efforts are sometimes puzzling. Following are actual quotes:

"I was appalled to learn that part of the fishing license revenues are used to prevent water pollution."

"They are basing this rule on studies of rats. Rats aren't people you know."

"They are the most anti-farmer DNR in the nation."

Commenting on why he voted against a new rule to keep a byproduct of a known carcinogen out of drinking water, one official said, "This rules shows that the DNR doesn't like people or the things that people do."

Every one of these comments was made by an elected state official in Madison.

How should public agency officials react to a legislative atmosphere like that described above? I believe we need to do all that we can to tell the story of science effectively.

All the points highlighted in this workshop are important, but in particular, I think that the "monitoring and estimation techniques" and "realistic expectations" tracks are keys to success. More research on sediment-loading prevention technology is important.

It is also critical to address environmental issues at the landscape scale. Data are a fundamental aspect of implementation and assessment of environmental policy and programs. Long-term monitoring must be part of this process.

Some key data and monitoring needs include the following:

1. Manure-spreading issues, like tile lines, and challenges with landscapes, like karst topography, need more analysis.
2. We also need systems approaches that do not overwhelm producers.
3. We need messages that are clear and connect to public health whenever possible.
4. Data must be local or clearly applied locally.
5. Data must be embraced by producers to be accepted.

If data can only be useful when embraced by producers, we must then ask the question: "Can modeling work?" I am not sure that models will be accepted by producers on the ground. I clearly understand that in-field data gathering is not practicable everywhere, but scientists, policymakers, and program managers also must recognize the reticence of many "on-the-ground"

folks to embrace models. In other words, new or enhanced mechanisms must be developed to explain the meaning of data and long-term monitoring to policymakers about why data and monitoring results are important.

Better messengers are needed as well once we know what the message is. Producers can be effective messengers. Work in Wisconsin with the state's Discovery Farms extends research and monitoring to actual working farms. Best management practices (BMPs) and laboratory models are tested on the ground, with private producers, and the results broadly publicized. This approach generates broad support in the farm community.

When communicating with the broader public, recent surveys of public attitudes towards environmental messengers suggest that health professionals are well respected and therefore should be engaged to carry these messages to the general public.

Farm groups must be actively engaged in public discourse on polluted-runoff challenges. The socioeconomic dimensions of these challenges also need to be part of any program or policy.

More and more evidence suggests that a landscape approach to managing environmental quality on agricultural land is needed. Particularly important is a systems orientation that is geographically scalable, includes key social factors, and can have a long-term context with real outcome assessments. While Wisconsin's Discovery Farms have produced good site-specific results, Discovery Watersheds may be more appropriate than Discovery Farms to tell a more complete picture.

Scientists, policymakers, and program managers must also recognize the fickle nature of the American public. Someone much wiser than me has observed that "instant gratification is not fast enough for most Americans." How should this fact be addressed in the design and implementation of policy?

Data needs are critical. Documenting the results of environmental management with clear, quantifiable data is a key means for explaining approaches and results to policymakers.

Also to be recognized is the social aspect of the environmental management equation, along with the realization that addressing the social side of these challenges will always be more difficult than dealing with the hard science side of the issues.

More education is needed regarding the concept of adaptive management. Although the approach has become a popular buzzword, the next step is to convince policymakers that adaptive management is an actual, effective management regime. Thereafter, opinion-makers must be engaged on the idea of adaptive public policy management. This may be a difficult hurdle to clear, however, because policymakers do not do well at setting realistic timeframes and providing the resources necessary to measuring and monitoring practices to determine if approaches are actually working. One way to achieve that result is to identify ways to empower the policymaker with the capacity to fail. Adaptive management ultimately is about educated experimentation, and until that fact is recognized and public policy practitioners are given the ability to try new approaches that may ultimately fail, the full potential of this management regime is not likely to be realized.

Some states, including Wisconsin, have begun to include adaptive management in legislation. A recent groundwater quantity law in Wisconsin not only created a new regulatory structure but also established a scientific Groundwater Advisory Council that has been empowered with the task of reporting back to the legislature by a date certain (December 31, 2007) regarding how well the law is working and what changes should be made to improve the law. This requirement was adopted specifically because legislators acknowledged that as more science became available that they should learn from that science and "adaptively manage" their statute to reflect that fact.

Another component of this adaptive management equation is collection of adequate monitoring data. How can on-going data collection and long-term monitoring be made "more sexy" for communicating with elected officials and the public? Funding for data collection and monitoring is an absolute essential whenever and wherever management regimes are developed and employed.

At the same time, shortcomings of this sort cannot be an excuse for inaction. As Franklin Delano Roosevelt once said, "We must take a method and try it. If it fails, admit it frankly and try another."

Only through an effective combination of data collection, monitoring, communication of results, and development of an adaptive management regime that offers policymakers the ability to fail can people learn from experimentation and move on. In a world where instant gratification is not fast enough, hard work will be necessary to achieve some high hopes instead of reverting to a climate of low expectations.

A Natural Resources Conservation Service perspective

William E. Puckett

The Conservation Effects Assessment Project (CEAP) thus far has exceeded the initial expectations of agency officials in terms of interest and partnership involvement. It has become a truly collaborative effort among a number of federal, state, university, and nongovernmental partners. Quantification of the environmental benefits and effects of conservation practices and program expenditures is of extraordinary interest to policymakers, legislators, and scientists. CEAP is a critical undertaking for the Natural Resources Conservation Service (NRCS), which confronts accountability and performance requirements daily. CEAP also is of vital importance to the science of conservation and future decisions that agricultural producers make on the land.

The tremendous interest in CEAP and the significant partnership involvement have translated into high levels of expectation in terms of outputs and products from the effort. It is extremely important that the CEAP partnership be realistic and sensitive to meeting those expectations. The science that undergirds conservation effort on agricultural land in this country must remain sound and defensible—no shortcuts or significant compromises can be tolerated. Demands for products, analysis, and conclusions or results need to remain reasonably well in line with the original purpose and intent of the project. CEAP will help answer many questions, but it will undoubtedly raise many more. Participants and onlookers must keep this in mind as the project evolves.

NRCS is a science-based agency. Throughout its history, leaders and employees have worked with many partners in the science and research community to develop the many conservation practices and practice standards in use today. Agency scientists are internationally recognized experts in soil survey and large-scale resource inventory and monitoring efforts, such as the National Resources Inventory (NRI). It is the task of agency officials to translate science into application on the landscape.

NRCS maintains approximately 160 conservation practice standards and more than 70 quality criteria for implementing those standards. The field office technical guide is the culmination of decades of scientific research on which the agency's creditability is based. Because of NRCS's conservation partners and the technical standards established by the agency, the "signature of conservation" is visible all across America.

Over the years, NRCS officials and partners in the conservation research and application communities have gathered much qualitative data and used those data to infer what the environmental impacts of conservation programs are on such resource concerns as water quality, water quantity, and soil quality. NRCS officials know, for example, that conservation practices and programs help to control soil erosion; this effect is assessed regularly through the NRI. But no methodology is in place for quantifying—measuring—the environmental benefits of conservation practices on the landscape. This is what CEAP is all about. CEAP is designed to help answer some very big and important questions, such as: What are the measured environmental benefits purchased by public conservation dollars? What's the bang for the buck in terms of environmental improvements brought about by conservation practices and programs?

Simply stated, CEAP is the manifestation of NRCS's efforts to understand the impact and interactions of conservation practices on the landscape. It is a test of the effectiveness and efficiency of applied conservation. It is a test of the science behind conservation practices and program efficacy and a determination of how to achieve more effective conservation on the land.

How do we quantify environmental gain? What is enough, acceptable, or a reasonable outcome for the public investment in conservation? What is a reasonable time frame or expectation for the desired effect to occur? Simply installing a conservation practice does not mean it is immediately functional. It takes time to reach optimum potential. The same practice may behave much differently from one resource region to the another depending upon climate, soils, landscape patterns, cropping systems, and management practices. What are some of the primary drivers of conservation activities? How do environmental goods and services and market-based conservation fit into the discussion and decision-making process? Farming is a business, and economic considerations must be evaluated.

How much should we rely upon modeling? What is the proper mixture of modeling and onsite monitoring and measurement? What should we or can we reliably measure and monitor? How much monitoring should be done?

What are the methodologies, technologies, and supporting infrastructure needed to support monitoring? Where should we measure and monitor? Are some resource concerns or areas more critical or easily monitored than others? What is the baseline or starting point for this effort? Do we take all land uses into account? These are all important questions.

At times, these questions seem overwhelming, but we must strive to answer them and others that arise during the CEAP effort. Most importantly, CEAP must be supported by the development of a sound and defensible science that is a necessary underpinning of all components of the effort.

Significant strides have been made in developing the CEAP partnership and the funding mechanisms required to support the first few years of the project. Far more work remains to be done, however, which underscores the importance of keeping CEAP on track as a productive enterprise.

Ongoing CEAP activities can be monitored at the following website: (http://www.nrcs.usda.gov/technical/NRI/CEAP).

A policymaking perspective

Anne Simmons

When making policy decisions, you often hear members of Congress talk about science-based policies. They often use the term "sound science" to describe what they think the executive branch, as well as their colleagues in the legislative branch, should be utilizing to make day-to-day decisions.

Congressman Stenholm always used examples of actions that dairies in the Bosque River watershed were told to take many years ago – before Texas Institute for Applied Environmental Research (TIAER) started working there – actions that didn't work. Now, I work for an accountant, so he's a "show me the numbers" kind of guy.

Congress tends to set broad policy goals and let agencies decide the best way to achieve those goals. Then, the members yell at the implementers if the laws do not achieve the goals in the way Congress thought they should.

It is extremely important to members of Congress—despite the fact that the funding provided does not always reflect that—to know that research is taking place and what the results and outcomes from that research might be. One important example that comes to mind is when folks like Tom Simpson from University of Maryland and Andrew Sharpley with Agricultural Research Service, as well as folks at TIAER in

Texas, got the attention of Congressman Stenholm and Congressman Gilchrest on the need to be just as concerned about the impact of phosphorus on water quality as nitrogen. A congressional briefing brought together individuals from the research community, with members of Congress, congressional staff, and key agency leaders, to discuss the state of research and findings from the field.

Ideally, I would like for the scientific community to be able to tell members of Congress and their staffs how to design performance-based programs, but concerns exist about going certain directions in agricultural conservation policymaking because of the many unknowns surrounding potential World Trade Organization (WTO) problems. The concern is that WTO does not always let member countries run programs in ways that make the most scientific or even common sense. The United States is criticized over direct payments in the commodity title not being tied to production, yet those payments were structured in that way to comply with WTO rules.

Perhaps the biggest issue remaining to be addressed in agricultural conservation is the question of targeting. Should programs target specific watersheds where significant environmental problems exist as a result of agricultural activities and concentrate financial and technical assistance in those watersheds? Is the nation wasting money by not targeting watershed by watershed? Or is there some benefit to putting conservation in more dispersed areas and to address issues that may not need specific targeting, such as wildlife habitat in some cases. Or do all of our areas for work need to be targeted?

It would be helpful also to have a good review of existing programs and the way agencies are carrying those programs out. Such looks at big-picture issues—if you can call them that—such as the aforementioned phosphorus example, would likewise help members of Congress decide if the focus and priorities of current programs need to be reevaluated.

The human factor is likewise an issue that raises a couple of important questions: First, what is the role of Natural Resources Conservation Service (NRCS) employees; should they be pushing paper in their offices or visiting producers? Second, are conservation districts and resource conservation and development councils living up to their roles and expectations? Sociologists must become more involved in delivering conservation assistance because success rests so much on those folks implementing programs in the field and the best ways to motivate and reach producers.

Related to both of the forgoing questions is the matter of funding. Additional institutions are needed to deliver funding—the role of nongovernmental organizations, the private sector, and state and local governments needs to continue and expand. NRCS cannot operate without these partners because Congress has not given the agency the needed resources. Just how we deal with NRCS workload issues remains uncertain. It will be interesting to work with the National Association of Conservation Districts, commodity and farm groups, and the agency to see if the will is there to make things better.

Some members of Congress also are asking whether we need to look at what needs to be done on the land. For example, we have spent consider-

able money on the Conservation Reserve Program and the Wetlands Reserve Program. Do we need to do better job of managing and maintaining those acres? What will be the driver to get more money for conservation? A WTO deal is unlikely. Budget cuts will likely be constraints. And the question remains: Who has the votes? If there is no safety net, does farmers' ability to do conservation suffer? I would be interested to learn if any economists have done any work on whether conservation work suffers when farmers' bottom lines suffer. There obviously are more questions than answers....

Thus far, support and demand for conservation spending has not waned. That does not mean, however, that we no longer need to monitor and evaluate what the public is getting for this investment. This is one reason why the Conservation Effects Assessment Project (CEAP) initiative is a priority.

Concluding remarks

Craig Cox

As the Managing Agricultural Landscapes for Environmental Quality workshop closed, I felt an exhilarating mix of excitement and anxiety—a familiar feeling to an obsessed fly fisher. It is the same feeling I get when I step into a trout stream and see a large trout sipping mayflies off the surface of the stream across from where I am standing. The excitement is there because that trout is feeding, and I have a chance to catch it. The anxiety is there because I may not be up to the challenge, not only because my skills may fall short, but also because a sudden gust of wind or an osprey flying upstream may send that trout into hiding for the rest of the day. Fly fishers learn early that unexpected events and luck can easily trump skill.

Some large trout swam through the workshop that led to the publishing of this book. Here are a few that impressed me most.

First, the contributions to the workshop and this book make it clear that we—the conservation community—know much, much more than we are using every day to manage agricultural landscapes. Science and professional experience are advancing rapidly. The potential to improve the effectiveness of our efforts to secure the natural resource base, rebuild ecosystems, and improve environmental quality by translating that knowledge and experience into more focused and strategic effort is great.

We must marshal resources to translate that knowledge into action. We must invest in the tools and, most important, in the continual education and training of the people who will use those tools to improve the environment in agricultural landscapes. As we advance science, we must redouble our efforts to spin off intermediate solutions even if those solutions are far from perfect. We must do this even at the expense of slowing our rate of scientific advance if budgets force us into such unpleasant choices.

Second, I sense in this project a nascent network of watershed- and landscape-scale experiments. Too often research and monitoring is occurring in one landscape while conservation program implementation is occurring in another. We have an enormous opportunity to bring research, monitoring, and implementation together in a network of "working watershed laboratories." Why not take 30 percent of the more than $4 billion annual budget for U.S. Department of Agriculture conservation programs and devote it to focused, place-based projects linked directly with research and monitoring of results? Why not create a network of projects intentionally designed to couple learning with doing? Wouldn't this be a way to create the robust, defensible platform that Peter Groffman, Phil Robertson, and Todd Walter referred to during the workshop and solve local problems with great meaning and value to our citizens? We could bring politics, science, and implementation together in a powerful mix to build momentum toward more effective management of our agricultural landscapes.

Third, I have been reminded of the need to step back and think about the realistic—perhaps plausible is more accurate—expectation of public policy. I have seen the desk of Anne Simmons, senior professional staffer of the U.S. House of Representatives Agriculture Committee, and I can assure you there is no button there that she can push to magically make all of what we have discussed here happen. I have not seen the desk of Todd Ambs at the Wisconsin Department of Natural Resources, but I suspect he has no such button either.

The best I think we can expect of public policy is to remove barriers and open doors. The best crafted policy is only as good (or bad) as the people who use it to take action on-the-ground. It is fundamentally up to us—individually and collectively—to seize whatever opportunities arise from successful efforts to reform public policy.

Fourth, I have been struck by the notion of "legacy effects" that came up several times during the workshop and in the chapters of this book. It is sobering to consider the extent to which decisions made 70 years ago are still limiting our options and constraining our future. It leads me to ask, what social and biophysical legacy are we creating today?

I ask that question with a growing sense of urgency. We are entering a period when the demands and stresses on our resource base, ecosystems, and environment are intensifying. Meeting demands for water and energy, in addition to food and fiber, will put serious pressure on our landscapes. The emergence of a bio-based economy holds both promise and peril for the future of those landscapes. And climate change will make all of our work more urgent and more challenging.

It is imperative, I think, that we drastically improve the effectiveness of conservation if we have any hope of achieving that sound and livable future that Stan Gregory says defines the ultimate objective of conservation and restoration.

Some months ago I heard someone define adaptive management as "getting a project started and managing the drift." That definition is very evocative to a fly fisher because that is what we do. We put our fly on the water and manage the drift downstream.

I encourage everyone with an interest in conservation to get our flies on the water, no matter what shape or size they are, and manage the drift as best we can. Even if we think our skills are not as good as we would like them to be.

Contributors

Authors

Arthur W. Allen
Wildlife Biologist, National Biological Service, U.S. Geological Survey, Fort Collins, Colorado

Todd Ambs
Administrator, Division of Water, Wisconsin Department of Natural Resources, Madison

Matthew Baker
Assistant Professor, Department of Watershed Sciences, Utah State University, Logan

L. Wesley Burger Jr.
Professor, Department of Wildlife and Fisheries, Mississippi State University, Mississippi State

Kathryn Boyer
Fisheries Biologist, West National Technology Support Center, Natural Resources Conservation Service, U.S. Department of Agriculture, Portland, Oregon

P.E. Cabot
Postdoctoral Research Associate, Department of Biological Systems Engineering, University of Wisconsin, Madison

Paul Capel
Research Chemist, U.S. Geological Survey, Minneapolis, Minnesota

Roger Claassen
Agricultural Economist, Economic Research Service, U.S. Department of Agriculture, Washington, D.C.

Craig Cox
Executive Director, Soil and Water Conservation Society, Ankeny, Iowa

Theo Dillaha
Program Director, Sustainable Agriculture and Natural Resource Management Collaborative Research Support Program, Virginia Tech, Blacksburg

Otto Doering
Professor, Department of Agricultural Economics, Purdue University, West Lafayette, Indiana

Mike Dosskey
Research Riparian Ecologist, National Agroforestry Center, Forest Service, U.S. Department of Agriculture, Lincoln, Nebraska

Jane Elliott
Research Scientist, Aquatic Ecosystem Impacts Research Branch, Environment Canada, Saskatoon, Saskatchewan

Stan Gregory
Professor, Department of Fisheries and Wildlife, Oregon State University, Corvallis

Peter Groffman
Senior Scientist, Institute of Ecosystem Studies, Millbrook, New York

Steve John
Executive Director, Agricultural Watershed Institute, Decatur, Illinois

Madhu Khanna
Professor, Department of Agricultural and Consumer Economics, University of Illinois, Urbana-Champaign

Catherine L. Kling
Professor, Department of Economics, Center for Agricultural and Rural Development, Iowa State University, Ames

Richard Lowrance
Ecologist, Southeast Watershed Research Laboratory, Agricultural Research Service, U.S. Department of Agriculture, Tifton, Georgia

Jim Miller
Research Scientist, Environmental Health Program, Agriculture and Agri-Food Canada, Lethbridge Research Centre, Lethbridge, Alberta

David J. Mulla
Professor, Department of Soil, Water, and Climate, University of Minnesota, St. Paul

P. Nowak
Chair, Academic Programs, Gaylord Nelson Institute for Environmental Studies, University of Wisconsin, Madison

F.J. Pierce
Director, Center for Precision Agricultural Systems, Washington State University, Prosser

William E. Puckett
Deputy Chief, Soil Survey and Resource Assessment, Natural Resources Conservation Service, U.S. Department of Agriculture, Washington, D.C.

C.L. Redman
Director, Global Institute of Sustainability, and Julie Ann Wrigley Professor, School of Human Evolution and Social Change, Arizona State University, Tempe

Kurt Riitters
Deputy Program Manager, Forestry Sciences Laboratory, Southern Research Station, Forest Service, U.S. Department of Agriculture, Research Triangle Park, North Carolina

G. Philip Robertson
Professor, Department of Crop & Soil Sciences and W.K. Kellogg Biological Station, Michigan State University, Hickory Corners

Anne Simmons
Senior Professional Staff, Agriculture Committee, U.S. House of Representatives, Washington, D.C.

Thomas W. Simpson
Professor and Coordinator, Chesapeake Bay Agricultural Programs, University of Maryland, College Park

K.M. Sylvester
Assistant Research Scientist, Inter-University Consortium for Political and Social Research, University of Michigan, Ann Arbor

Mark Tomer
Soil Scientist, National Soil Tilth Laboratory, Agricultural Research Service, U.S. Department of Agriculture, Ames, Iowa

Todd Walter
Assistant Professor, Department of Biological and Environmental Engineering, Cornell University, Ithaca, New York

Mary C. Watzin
Professor, Rubenstein School of Environment and Natural Resources, University of Vermont, Burlington

Sarah Weammert
Faculty Extension Assistant, BMP Project Leader, University of Maryland, College Park

John Wiens
Lead Scientist, The Nature Conservancy, Arlington, Virginia

Wanhong Yang
Associate Professor, Department of Geography, University of Guelph, Guelph, Ontario

CEAP Technical Committee

Jerry Bernard
Natural Resources Conservation Service

Jan Boll
University of Idaho

Dale Bucks
Agricultural Research Service

Mike Burkart
Agricultural Research Service

Richard Butts
Agriculture and Agri-Food Canada

Craig Cox
Soil and Water Conservation Society

Tom Drewes
Natural Resources Conservation Service

Lisa Duriancik
Cooperative State Research, Education and Extension Service

Tom Franklin
The Wildlife Society

Bill Gburek
Agricultural Research Service

Brook Harker
Agriculture and Agri-Food Canada

Robert Kellogg
Natural Resources Conservation Service

Cathy Kling
Iowa State University

Scott Knight
Agricultural Research Service

Tom Lederer
Farm Service Agency

Martin Locke
Agricultural Research Service

Mike O'Neill
Cooperative State Research, Education and Extension Service

Roberta Parry
Environmental Protection Agency

Charles Rewa
Natural Resources Conservation Service

Clarence Richardson
Agricultural Research Service

Carlos Rodriguez
Agricultural Research Service

Mary Ann Rozum
Cooperative State Research, Education and Extension Service

Max Schnepf
Soil and Water Conservation Society

Andrew Simon
Agricultural Research Service

Janice Ward
U.S. Geological Survey

Mark Weltz
Agricultural Research Service

Program Committee

Hannibal Bolton
U.S. Fish and Wildlife Service

Mike Burkart
Agricultural Research Service

Roger Claassen
Economic Research Service

Craig Cox
Soil and Water Conservation Society

Mike Dosskey
National Agroforestry Center

Katie Flahive
U.S. Environmental Protection Agency

Karen Flournoy
U.S. Environmental Protection Agency

Stan Gregory
Oregon State University

Peter Groffman (chairman)
Institute of Ecosystem Studies

Brook Harker
Agriculture and Agri-Food Canada

Jon Haufler
Ecosystem Management Research Institute

Skip Hyberg
Farm Service Agency

Cathy Kling
Iowa State University

Martin Locke
Agricultural Research Service

Lee Norfleet
Natural Resources Conservation Service

Pete Nowak
University of Wisconsin-Madison

Mike O'Neill
Cooperative State Research, Education and Extension Service

Larry Payne
U.S. Forest Service

Fran Pierce
Washington State University

Peter Richards
Heidelberg College

Max Schnepf
Soil and Water Conservation Society

Karen Solari
U.S. Forest Service

Fritz Steiner
University of Texas

Dave Walker
U.S. Fish and Wildlife Service

Janice Ward
U.S. Geological Survey

Index